The Author Kenji Kawakami began his creative career in the Sixties writing scripts for animated TV series, and went on to arrange media events, becoming an avid opponent of such Japanese institutions as karaoke. After designing the Tokyo Bicycle Museum in the mid-Eighties, he became editor of a popular home shopping magazine, in which the concept of Chindogu was first born. The author of four books of unuseless inventions, and founder of the 10,000-strong International Chindogu Society, he is famous throughout Japan for his bizarre gadgets and his tireless work for the better understanding and appreciation of the tenets of Chindogu.

The Translator Dan Papia gave up a successful career as a financial journalist in 1991 to focus on more serious pursuits, and almost immediately created the so-called 'eye paddle' — a ball painted to look like an eye attached to an optician's chart paddle. He gave the Tokyo Dome its official name ('Big Egg'), and introduced Chindogu to the world outside Japan. He currently runs a video distribution business, writes a media column for the *Mainichi* newspaper and a humorous column for the *Tokyo Journal* and appears irregularly on Japanese TV.

The Editor Former *Punch* restaurant critic and *Sunday Telegraph* columnist, Hugh Fearnley-Whittingstall is a writer and broadcaster whose primary journalistic interest has been food. It was pursuit of the deadly Japanese delicacy fugu (or puffer fish) that first brought him to Japan. He has produced several documentaries for television, including *The Maggot Mogul,* about a scheme to build the world's biggest maggot farm. Recently he made his debut as a television presenter on Channel 4, as the resourceful scavenger-gourmet in *A Cook on the Wild Side.*

101 UNUSELESS JAPANESE INVENTIONS

The Art of *Chindogu*

Kenji Kawakami

Translated by and additional text by Dan Papia
Edited by Hugh Fearnley-Whittingstall

HarperCollinsPublishers
77–85 Fulham Palace Road,
Hammersmith, London W6 8JB

A Paperback Original 1995
5 7 9 8 6 4

The author is the Founder and President
of the International Chindogu Society

A catalogue record for this book
is available from the British Library

ISBN 0 00 647932 4

Set in Helvetica

Printed in Italy by Printer Trento Srl

Editor's Foreword

Genius has most satisfactorily been described as the infinite capacity to take pains. By this reckoning the creators of the carefully crafted Chindogu in this book must rank as genius every one. The fact that they often create more pains than they take adds an extra dimension to their brilliance.

The original genius and founder of Chindogu is Kenji Kawakami, designer, anarchist, pathological mail-order enthusiast, and author of this book. I met him in December 1993, on the last evening of a ten-day trip to Japan while writing some articles for the *Sunday Telegraph.* Kawakami's work was to have been the subject of one of those articles. But when he showed me the half-dozen Chindogu that were scattered about his profoundly untidy office (among them the Eye Drop Glasses and the Panorama Camera), and then produced three books with photographs of hundreds more mind-unbending inventions, I realised I had met someone who could truly speak to the world. An article, disposable and ephemeral, would not be enough.

As he demonstrated the solemn function of the Fish Face Cover and the Telephone Dumbell, his wide-eyed earnestness collapsed periodically into infectious giggling. And I began to glimpse the universal charm of these Chindogu (a word which is, you should note, both plural and singular – very singular). Designed to solve many of the niggling little problems of modern life, at home, at work, at leisure (and, incidentally, while commuting between the three) they have a tendency to fail completely – but also heroically, magnificently and beautifully.

For me, the best thing about Chindogu is that they are real: all the inventions that appear in this book have actually been made, painstakingly and skilfully, by Kawakami and fellow Chindogu enthusiasts in Japan. They are also surreal – in even contemplating their use you enter a new and uncharted dimension of human endeavour. And Chindogu are full of stimulating contradictions. They are at once gentle and subversive, anarchic and mundane, ingenious and stupid. Are they art? It probably doesn't matter. But it definitely matters that it probably doesn't matter.

At the end of my intoxicating first encounter with Chindogu, I enthusiastically agreed to help Kawakami bring his work to the attention of the English speaking world. Almost two years later, this book is the result. I feel that my contribution as editor may not at times have been entirely unuseless. But it has certainly been fun.

Hugh Fearnley-Whittingstall
May 1995

珍道具

INTRODUCTION

There are those who date the dawn of our species at the invention of the first tool. Perhaps mankind can be said to have originated the moment that a certain hairless and ambitious ape threw a stone at a deer and turned the tedious task of chasing and killing into a far less loathsome chore.

The primitive tools that have since punctuated the process of our so-called civilization - at least according to the history books - were all to some degree more advanced than the previous ones. But we can suppose that for every inspired notion that advanced the human situation, there must have been at least as many duds; ideas for improvement that didn't quite work. Another hairless ape, for example, maybe the brother-in-law of the first tool-making human, might have grown tired of eating deer and envisioned a huge hill-sized boulder flying through the air and downing a nice thick woolly mammoth, only to finally decide that the rock would be too hard to throw and the whole operation certainly more trouble than it was worth. You'd have to roll it up a mountain, wait for the right chance, hope the beast wouldn't move...

It is this second sort of idea – just as noble but fruitless and until now less appreciated – that the art of Chindogu was conceived to celebrate. The Japanese word Chindogu literally means an odd or distorted tool – a faithful representation of a plan that doesn't quite cut the mustard.

The successful Chindoguist approaches his subject in much the same way that a serious inventor would: searching for an aspect of life that could somehow be rendered more convenient and concocting a method for making it so. Like the inventor, he discards those notions that clearly miss the mark, but unlike the inventor, he also abandons those ideas that will obviously work. The Chindoguist latches onto and builds a prototype of the best idea he can come up with that looks good at the onset but on closer examination isn't. Having tested and verified that it indeed wasn't worth the effort, the creator of the Chindogu will then congratulate himself on having successfully produced an almost useful implement.

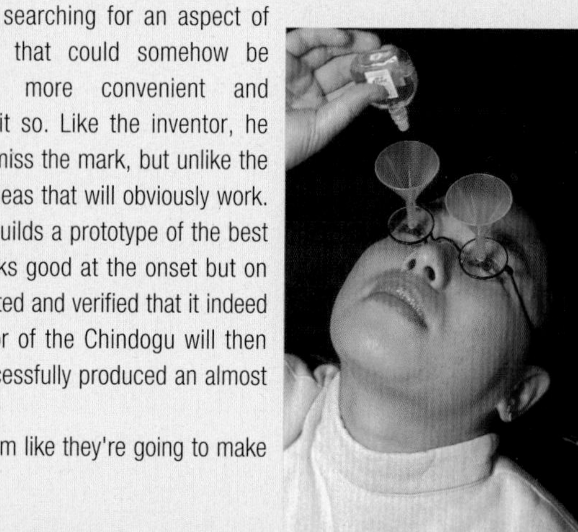

Chindogu are inventions that seem like they're going to make

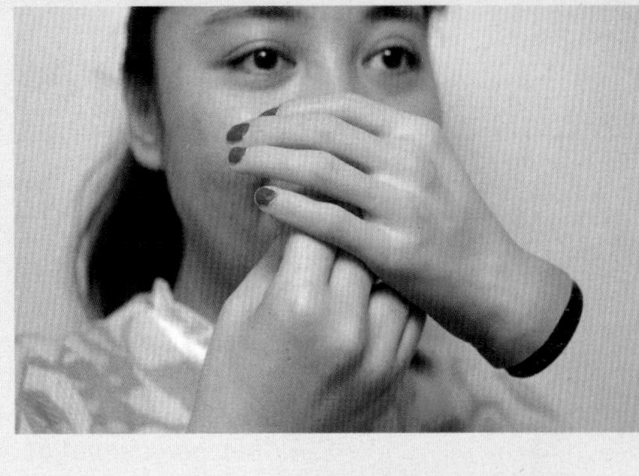

life a lot easier, but don't. Unlike joke presents built specifically to shock or amuse, Chindogu are products that we believe we want - if not need - the minute we see them. They are gadgets that promise to give us something, and it is only at second or third glance that we realize that their gift is undone by that which they take away.

They are funny because they are paradoxical and they are funny because they fail, but Chindogu are not designed to be stupid for the mere sake of stupidity. To believe this would be a grave error. They are exactly what we've always longed for without actually realizing it; they are inventions that do a job, satisfy a need; they are everything it takes to bridge the distance between what we are and what we could be...almost.

Unlike the hungry simian with his unfulfilled craving for mammoth meat, we children of the prosperous twentieth century have the luxury of being able to build that which we can't really use. Separated from the realm of practicality and profits, from the constraints of utilitarian application, they can take us into a new and spectacular world of human invention whenever we are lonely, angry or depressed. All we have to do is adhere to the ten tenets of Chindogu to insure that the purity and integrity of the discipline is maintained, and the process of encouraging and preserving the almost inspired and nearly brilliant becomes an instinctive way of life. We can scarcely look at a coffee cup without wondering whether another handle would make it twice as easy to pick up (and, if so, shouldn't ten more increase the convenience tenfold?). Chindogu lends an ever present quality of anarchy to even our most common concerns and a new dimension to our earthly existence.

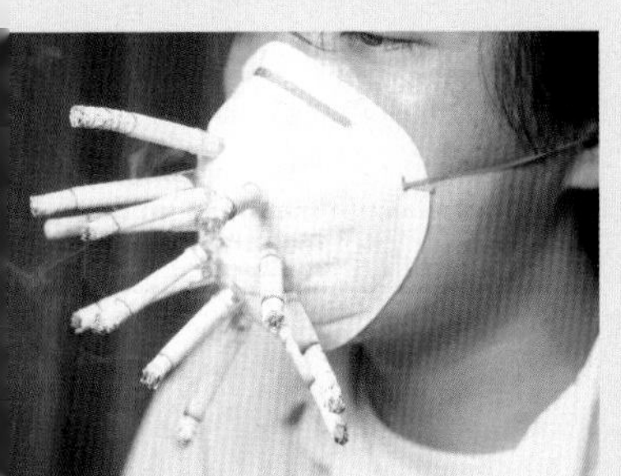

In simple 1990s' terms, the practice of Chindogu is like a Jurassic Park for those less fortunate of man's labour-saving dreams and schemes. The International Chindogu Society protects and nurtures those ideas that technological evolution would otherwise doom to extinction so that we can, at our leisure, observe them, admire them and enjoy them.

We hope this book will help you, too, to take part in and take pleasure from this new and exciting form of non-verbal communication.

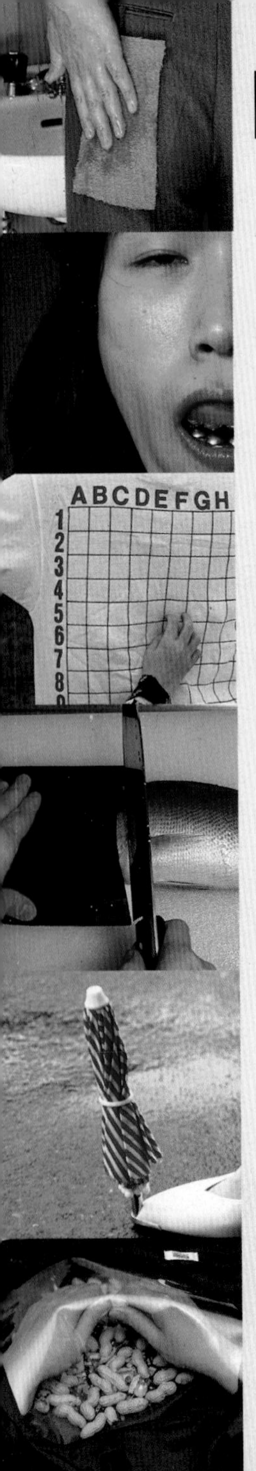

The Ten Tenets of Chindogu

Every Chindogu is an almost useless object, but not every almost useless object is a Chindogu. In order to transcend the realms of the merely almost useless, and join the ranks of the *really* almost useless, certain vital criteria must be met. It is these criteria, a set of ten vital tenets, that define the gentle art and philosophy of Chindogu. Here they are:

1. A Chindogu cannot be for real use

It is fundamental to the spirit of Chindogu that inventions claiming Chindogu status must be, from a practical point of view, (almost) completely useless. If you invent something which turns out to be so handy that you use it all the time, then you have failed to make a Chindogu. Try the Patent Office.

2. A Chindogu must exist

You're not allowed to use a Chindogu, but it must be made. You have to be able to hold it in your hand and think 'I can actually imagine someone using this. Almost.' In order to be useless, it must first be.

3. Inherent in every Chindogu is the spirit of anarchy

Chindogu are man-made objects that have broken free from the chains of usefulness. They represent freedom of thought and action: the freedom to challenge the suffocating historical dominance of conservative utility; the freedom to be (almost) useless.

4. Chindogu are tools for everyday life

Chindogu are a form of nonverbal communication understandable to everyone, everywhere. Specialised or technical inventions, like a three-handled sprocket loosener for drainpipes centred between two under-the-sink cabinet doors (the uselessness of which will only be appreciated by plumbers), do not count.

5. Chindogu are not for sale

Chindogu are not tradable commodities. If you accept money for one you surrender your purity. They must not even be sold as a joke.

6. Humour must not be the sole reason for creating a Chindogu

The creation of Chindogu is fundamentally a problem-solving activity. Humour is simply the by-product of finding an elaborate or unconventional solution to a problem that may not have been that pressing to begin with.

7. Chindogu is not propaganda

Chindogu are innocent. They are made to be used, even though they cannot be used. They should not be created as a perverse or ironic comment on the sorry state of mankind.

8. Chindogu are never taboo

The International Chindogu Society has established certain standards of social decency. Cheap sexual innuendo, humour of a vulgar nature, and sick or cruel jokes that debase the sanctity of living things are not allowed.

9. Chindogu cannot be patented

Chindogu are offerings to the rest of the world – they are not therefore ideas to be copyrighted, patented, collected and owned. As they say in Spain, *mi Chindogu es tu Chindogu.*

10. Chindogu are without prejudice

Chindogu must never favour one race or religion over another. Young and old, male and female, rich and poor – all should have a free and equal chance to enjoy each and every Chindogu.

The Lipstick Seal

✷ *For ladylike smudgeless sipping*

Ladies (and other lipstick wearers), your beverage consumption need no longer be marred by the ever present fear of leaving unsightly lip-prints on your drinking vessel. The lipstick seal is an adhesive stick-on transparent tab that will banish this inadvertent discourtesy from your life forever. Just peel off and throw away when the drinking hour is over, and the lip-contact portion of your glass or mug will still be pristine clean.

Lady's Coffee Mug

✷ *The alternative lip-smudge avoidance device*

Should the lipstick seal prove user unfriendly, or to avoid the necessity of carrying spares, the Lady's Coffee Mug will solve the coffee-break lip-print problem once and for all.

Based on the time-honoured principle of camouflage, the sipping area of the mug, both inside and out, is coloured lipstick red, rendering those lip prints invisible to social score-settlers and etiquette terrorists. And no used lipstick seals to clutter up the ashtrays.

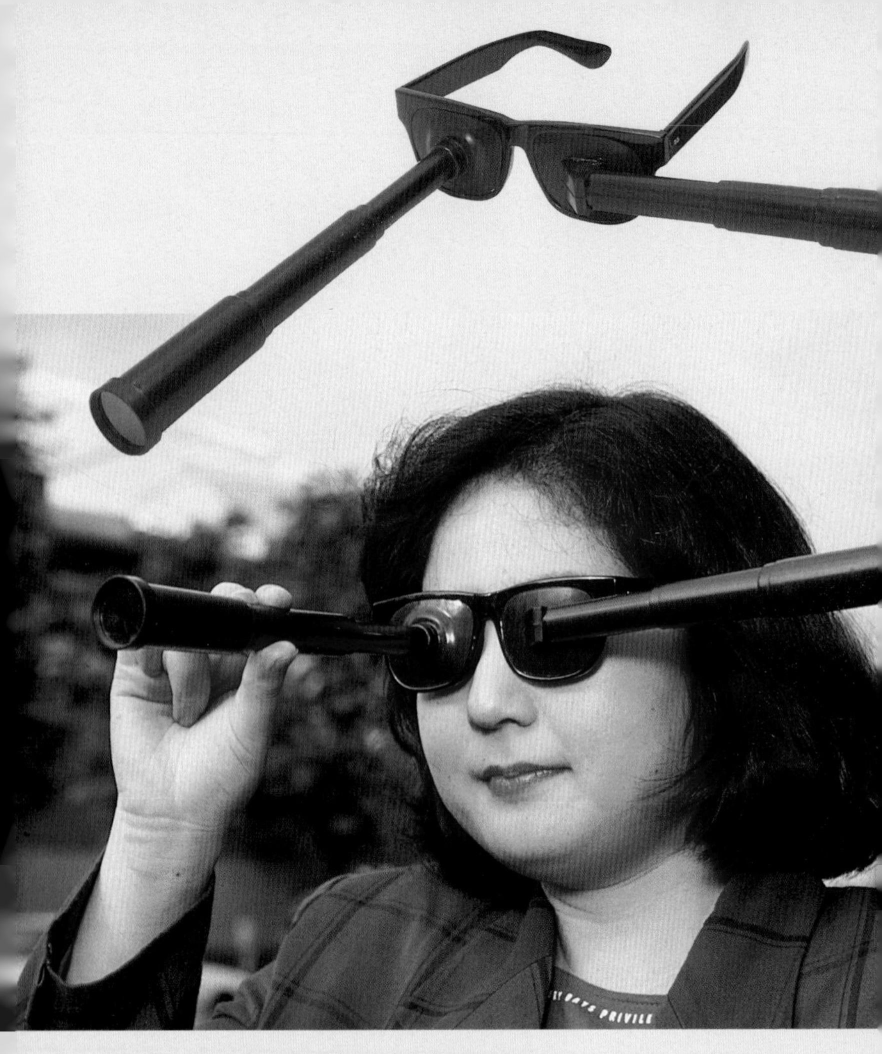

The Forward-Backward Glasses

✳ *For simultaneous front and rear vision*

Recognise your friends before they surprise you from behind. Spot the long arm of the law creeping up on you. Anticipate those pushy pedestrians. With the left eye looking forward, the right eye back, you need never miss a trick.

Unfortunately human ocular evolution has not kept pace with such ingenious technology, and wearers of the Forward-Backward Glasses will struggle to achieve simultaneous front and rear vision. But don't be downhearted. Keep a pair in the family, and your great, great, great, great, great, great, great, great, great, great, great (and then some) grandchildren may feel the benefit.

Duster Slippers For Cats

✷ For feline assistance with tedious housework

Now the most boring job around the house becomes hours of fun. Not for you, but for your cat! With these dust-dislodging foot socks, cats can play their part in easing the pressure of domestic chores. Lazy cats are of course much less productive than excitable ones, but this problem may be overcome if you introduce a dog into the house.

The Telephone Dumbbell

✱ *Increases fitness and reduces phone bills*

Lazy kids never off the phone? Discourage lengthy phone calls and turn fat to muscle with the telephone dumbbell. The five-kilo standard model means that most calls will run under three minutes, but serious scrimpers may be interested in the ten-kilo phone, which brings average talk time down to a super-brisk, 'hello-goodbye' thirty seconds.

In the public domain, the Dumbbell Phonebox reduces waiting time and boosts national fitness.

カード公衆電話
カード
テレホンカードと
100円・10円硬貨が使えます
カードの残りは
度数です
100円 10円
100円はおつりがでません

P A R S
まさか のはずが

Newspaper Bath Packs

✷ *Vinyl waterproofing for broadsheet newspapers*

Rubber ducks may be fun but they won't help you get ahead in the morning. Scientists have shown that a good hot bath stimulates alpha waves in the brain. So, for optimum use of bathtime, get your body clean and your brain ready for the day — read the paper in the bath.

One vinyl pack encloses four pages of news, counting the front and back. So, for a thirty-two-page section, only eight recyclable packages are required. And they don't just keep the water out, they keep the newspaper ink in. So no more cloudy grey papier-mâché baths.

The Noodle Eater's Hair Guard

✱ *Helps rapid lunchers protect expensive coiffeur*

The Samurai hairstyle may have been an ancient and artful solution to the problem of maintaining follicular hygiene at mealtimes. Fashion has changed, and this age-old problem has returned. The modern solution is the Noodle Eater's Hair Guard. In stylish candy pink moulded plastic it fits neatly into briefcase or handbag.

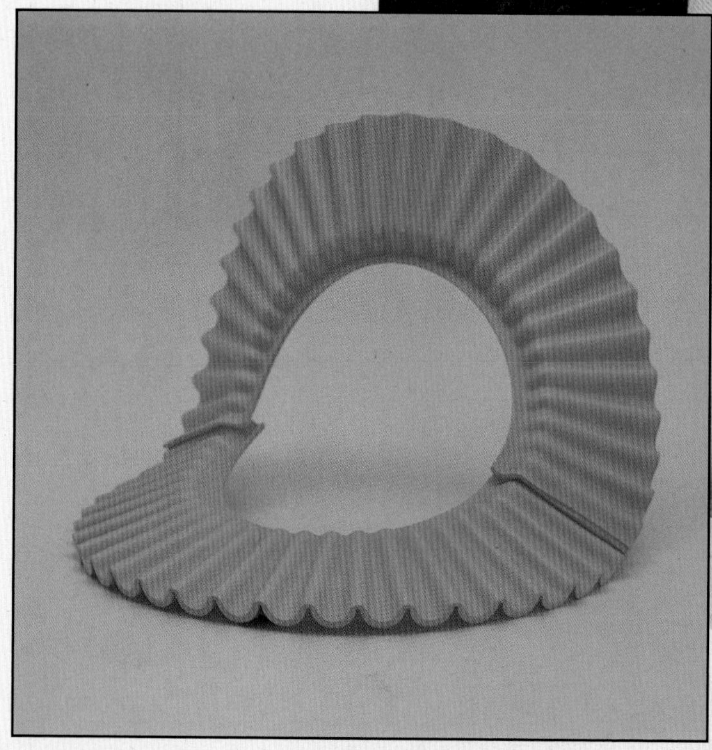

Housewife's Snooze Aid

✷Wave goodbye to housework fatigue

The exhausted housewife merits the occasional lie-in. But such indulgence can jeopardise marital bliss. How, for example, can the snoozing wife wave goodbye to her husband as he leaves for work? The answer is the Housewife's Snooze Aid, an automated waving hand, attached to an alarm clock, and timed to give a cheery wave just as hubby puts his head round the door to announce his departure. Hang the 'welcome back' sign underneath, and she can get away with sleeping the whole day through.

The Earring Safety Net

✷ *Wherever you go, your earrings come too*

The loss of an earring can be sad, expensive, inconvenient – even incriminating (depending where the earring turns up). Don't let these negative forces spoil your day. Travel everywhere with your earring safety nets firmly in place – and don't forget to check them regularly.

23-6

Full Body Umbrella

✷ *For day-long all-over dryness*

The umbrella, invented over 200 years ago, was a good start in the technological battle to stay dry in the rain. But total victory has had to wait until now. Protected by a 360° sheet of clear vinyl, users of the full body umbrella are safe from even those sneaky, low-flying horizontal rain drops. And other pedestrians will appreciate the lack of dangerous spikes at eye level. Watch out for small dogs though.

Stretch Spoon

✷ *Whenever that extra reach is needed*

The household spoon has evolved to its popular modern shape to serve most of our modern needs. But the one thing it just can't do is reach the bottom tenth of the average coffee jar. All too often the coffee maker is forced to improvise, and shake the last few spoonfuls from the jar straight into the cup. This slovenly practice leads to unpredictable beverage strength and reduced coffee break enjoyment. The Stretch Spoon brings relief. With its expanding telescopic handle it transforms from the regulation eleven centimetres to an impressive twenty-three.

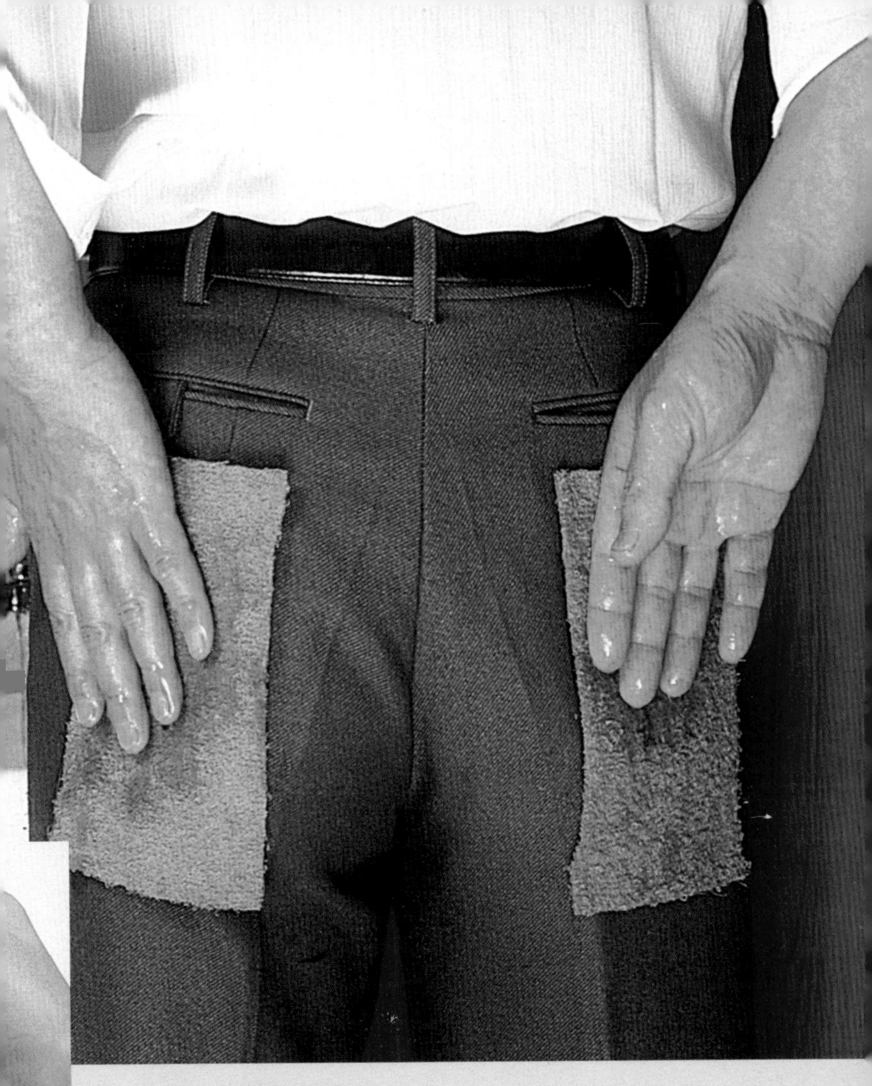

Pant Hankies

✱ *A better life for impulsive hand wipers*

Too many of us are opting for a convenient but ultimately unsatisfactory piece of material on which to wipe wet hands — the seat of our pants. The upside, of course, is dry hands. But is it worth the downside — wet pants?

Of course, old habits die hard — so if you can't change the man, change the pants. Two thin strips of velcro on the seat of pants or skirt allow the wearing of detachable Pant Hankies. No need to curb the wiping instinct, and no more humiliating rear end rashes.

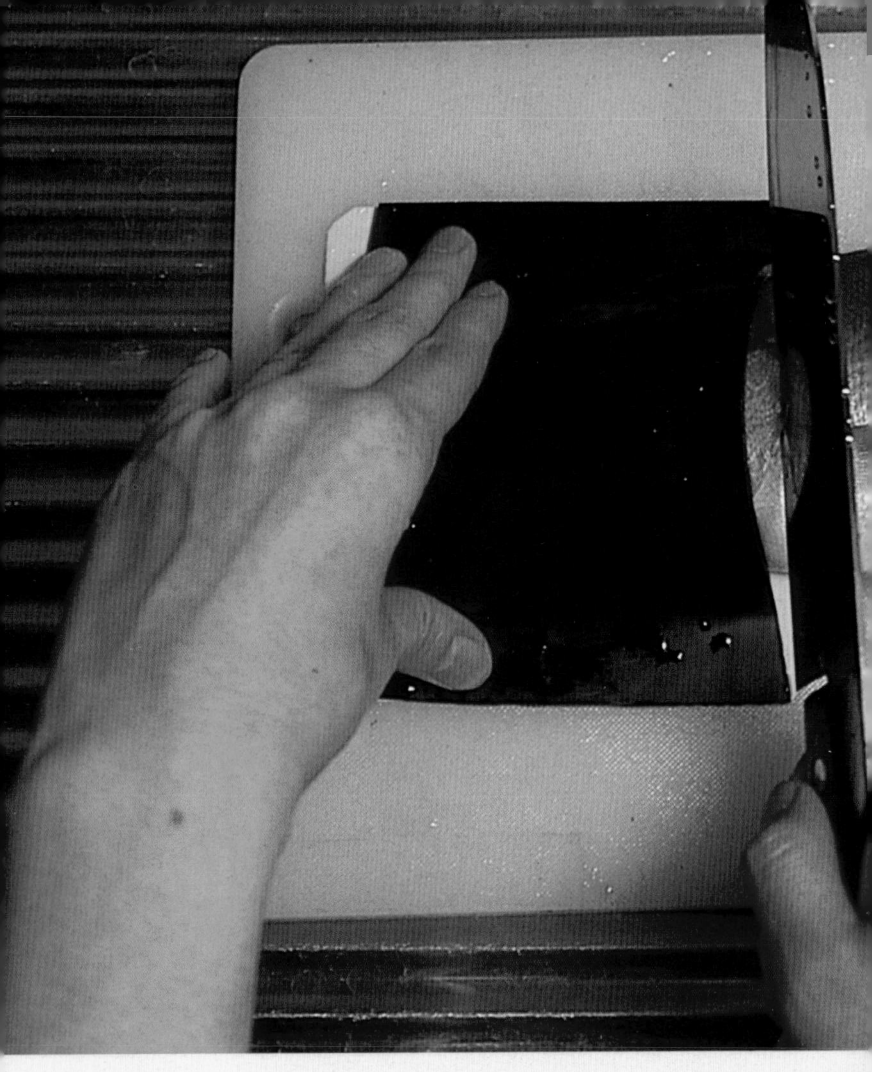

Fish Face Cover

✷ Helps get the fish cut up with minimum emotional trauma

Too many modern home cooks feel unable to look a dead fish in the eye. This has led to some nasty kitchen accidents, as the cook continues to slice with averted gaze. This in turn has given rise to the popularity of takeaway sushi and breadcrumbed fish fingers (you can buy five and shake them by the hand, but you still don't have to look them in the eye). In order that the art of fish cookery should not be lost forever, Chindogu brings you the Fish Face Cover. Slip it over the fish's head just prior to decapitation and avoid that reproachful stare.

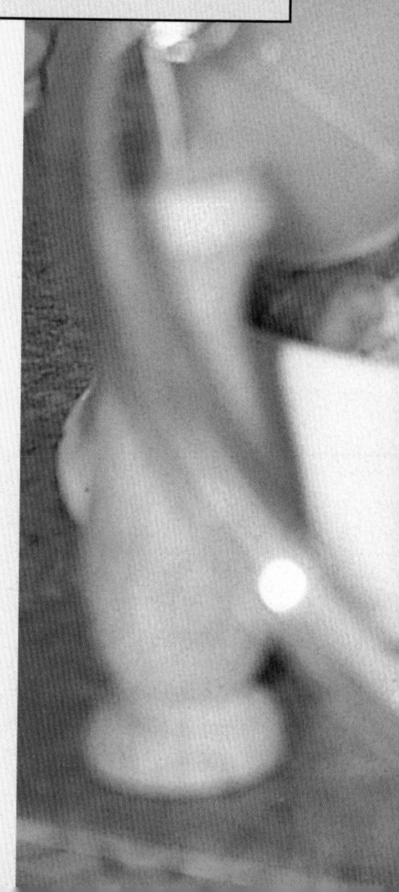

Cat Tongue Soother

✱ *Food temperature control for cats*

Anyone who feeds their cat from the table will be aware of the feline aversion to hot food. In Japan this phenomenon is so familiar that people who have to blow on their noodles are said to have a 'cat tongue'. But sadly, owing to an evolutionary oversight, real cats can't blow on their food to cool it down.

They can now! This simple device allows cats to chill their chow with paw power. Now you can serve up piping hot food to your pet without fear of a visit from the Cat Protection League.

珍道具

Cranium Grabbing Glasses

✷ *For short-sighted sideways snooozers*

As anyone who requires corrective eyewear will tell you, the problem with wearing spectacles is not the lenses, it's the arms. And never more so than when you nod off during bedtime reading, or horizontal telly-watching. If you're not instantly woken by a jab behind the ear, then chances are you'll wake up with a one-armed pair of specs.

Banish such bother and breakages, with the sleep-safe Cranium Grabbing Glasses. Pilot wearers have complained of being laughed at. But are they really any less flattering? When the whole world has converted, we may decide not.

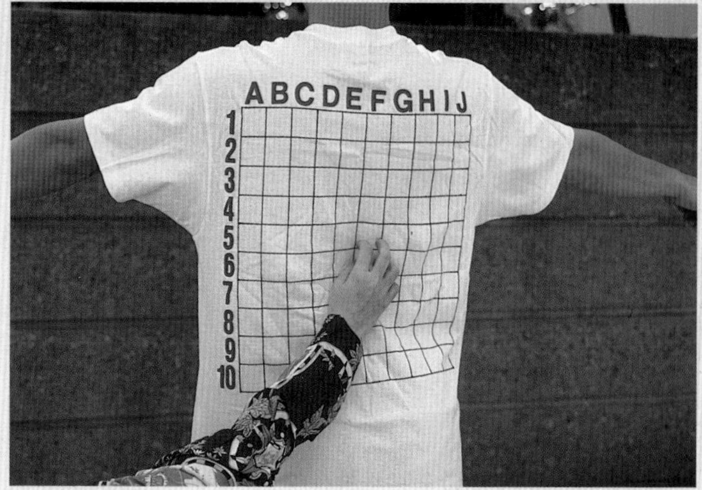

Back Scratcher's T-Shirt

✷ *The fast and logical solution to infernal itching*

The friend (or partner) who offers to scratch your back is a friend (or partner) indeed. Except it all goes horribly wrong when they just can't seem to locate the maddening itch. For those who are fed up of saying, 'left a bit ...up a bit ...right a bit ...damn!' comes a very special t-shirt, complete with Battleships style, itch-locater grid. The scratchee is also equipped with a hand-held miniature corresponding grid-map, for accurate communication. So when the scratcher says, 'I'm scratching F5,' the scratchee can say, 'try G7'.

Handy Knife and Handy Cutting Board

✱ *For hands-on food preparation*

For maximum control in a small space, and that invaluable feeling that one is completely in touch with one's produce, this complimentary pair of gloves constitutes a miniature portable food preparation system – both work surface and blade. Once you get used to them, you will hardly know where your body ends and your utensils begin. Not recommended for those who like to pick and nibble while they cook.

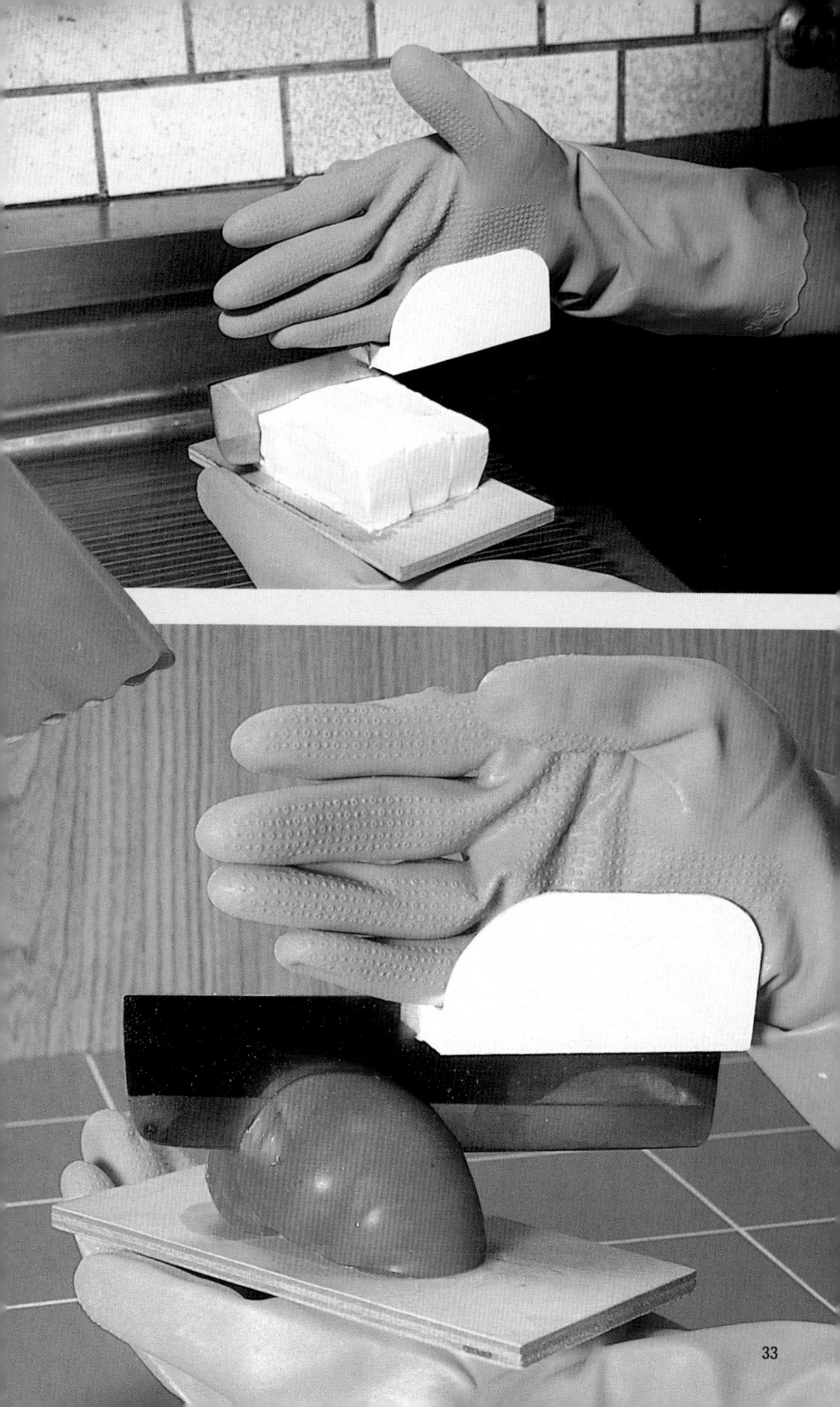

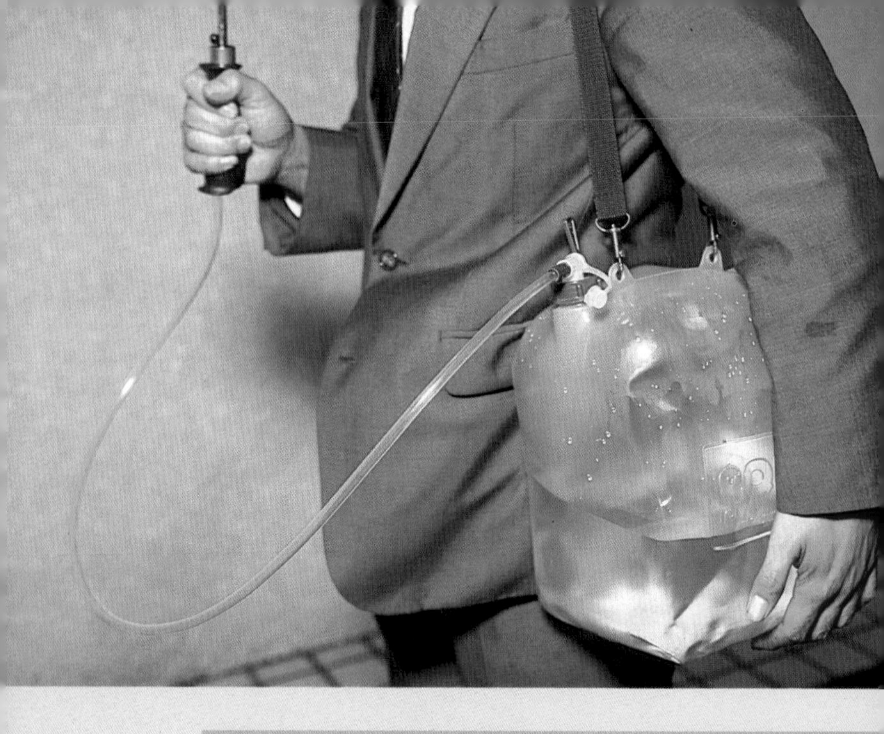

Personal Rain Saver

✷ ***Every drop that falls is yours to keep***

No one has any illusions about the value of water as a natural resource. Yet all of us watch hundreds of gallons washing off our umbrellas and into the gutters every year. With the Personal Rain Saver you not only stay dry, you can collect your own share of natural precipitation in a shoulder-harnessed tank.

As an alternative to mass assembly, the Personal Rain Saver provides a pro-planet pastime for peaceful protestors to engage in on their own (weather permitting).

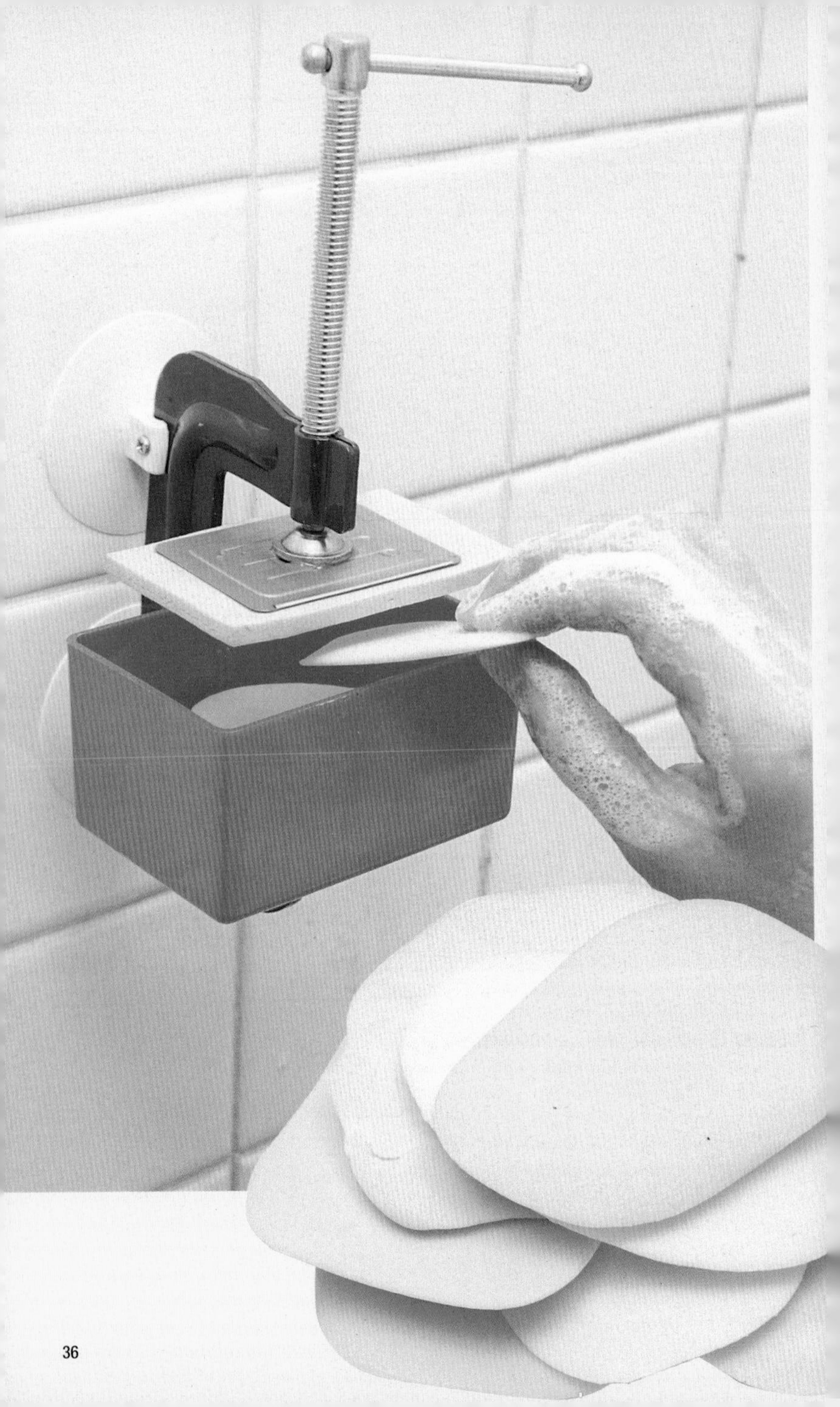

Soap Recycler

✱ *The answer to sud's law*

When a bar of soap is finally reduced to the little slippery slither that you just can't keep hold of, it becomes the scrimper's worst nightmare. You can't bear to throw it away, but you can't do a thing with it either. That's sud's law.

So save your slivers, until you have enough to fill your soap saver. Then turn the handle, and press the useless remnants into an all new bar of soap — quite possibly multi-coloured, multi-scented, and for all skin types.

Note: disturbing rumours have reached us that such a product is being marketed commercially in the US. Its status as Chindogu is therefore under review.

Step Sandals

✷ *The ultimate high heels*

From earliest times, man has reached for the skies: hence the Tower of Babel, the hot air ballon and, at a more mundane domestic level, the step ladder. The latter is what most of us use to get a little lift in life, but it has its drawbacks – most obviously, when you move, it doesn't.

Step forward the step sandals, the latest in extendable tripod-fitted footwear. They provide artificial elevation with safe equilibrium. The domestic advantages are many: high dusting, ceiling painting, top bookshelf access. But you can also walk tall outside the home: see what the rest are missing when crowds gather and you're late on the scene. It also allows children to see the grown-up's point of view.

谷合

Shopper's Umbrella

✷ *Increases purchase power in a downpour*

For over a century now the umbrella has kept the rain off human heads, but at a price – one human hand. The committal of a full half of the carrying limbs to the business of holding the umbrella drastically reduces the ability of shoppers to carry the fruits of their acquisitional frenzy.

That's why the sturdy Shopper's Umbrella is fitted with eight hanging hooks attached to the canopy-supporting rods. These can be used to house and carry bags of all sorts, which will join you in the rain-free zone. Novices should beware of two small potential hazards: a poorly balanced umbrella and impaired vision.

トイレに流しても安心です!
EAUDOUCE
水に流せるティシュ
EAUDOUCE
オーデュス
NSSHNBO
EAUDOUCE

Automated Noodle Cooler

✷ *Further assistance for the noodle-eater-in-a-hurry*

It is fundamental to Oriental cuisine that noodle soups are always served at a temperature at which they are infeasibly hot. The admirable reason is consumer choice; the diner can decide when his or her noodles have cooled sufficiently to be deemed palatable. But today, when time is money and money is scarce, not everyone can afford to wait around for noodles to cool. Frantic blowing is both undignified and limited in its effectiveness. The Automated Noodle Cooler is the answer. A compact fan, it fits tidily towards the holding end of one of a pair of chopsticks.

The Noodle Cooler can also be adapted to fit spoons and forks, and hence combat the problem of too-hot food as it arises in other cultures.

Butterstick

✷ *Why dirty a knife?*

Somewhere beyond the lipstick, gluestick and the stick deodorant comes the latest in stick-type-applicator-technology: the butterstick. Its manner of use is self-evident. Besides toast, it is useful for buttering corn on the cob and crumpets; less useful, unfortunately, for buttering rice and peas. Note also that, when cold from the fridge, the butterstick may have a tendency to 'flay' soft bread.

Soft Drink Holder

✷ *Even people with busy hands deserve refreshment*

There are other devices that go by this name, but they are undoubtedly inferior. Why? Because they hold your drink not when you need it, only when you don't. The Chindogu Soft Drink Holder (plus extra-long Chindogu straw) gets drink to your mouth while your hands are more gainfully employed.

Shoe Tabs

✷ *Facilitate instant shoe removal*

Shoes that are reluctant to leave your feet can be a nuisance. Shoes that are very reluctant to leave your feet can be dangerous. Even if you succeed in prising off one shoe by using the heel of another, you have to put the first shoe back on again to get the other one off. And when you're hopping around, one shoe off, one shoe on, accidents can happen. Shoe tabs make all the difference. The shoeless heel can safely and surely push on the tab to remove the shoe from the still-shod foot. But beware of striding confidently through a crowd, and finding yourself suddenly shoeless.

Sweetheart's Training Arm

✷ *A PDA (Public Display of Affection) confidence developer*

It's a tough life, these days, for a nervous or novice sweetheart – the kind of person who looks around at the young couples of today, strolling arm in arm and hand in hand, and thinks, 'I admire that bold statement of affection. But I'd never be able to do it myself.'

To the aid of such unfortunates comes this indispensible training device, a dummy arm. It attaches to a jacket or coat just next to the real arm. You can hold onto this simulated appendage in the early stages of courting, without the worry of sweaty palms, inappropriate pressure, or when to disengage.

Both male and female dummy arms are available, but it is advised that only the less confident partner of the couple attempts to train with the dummy arm. Otherwise a bizarre situation may arise, of doubtful benefit, in which one dummy arm is linked affectionately with another.

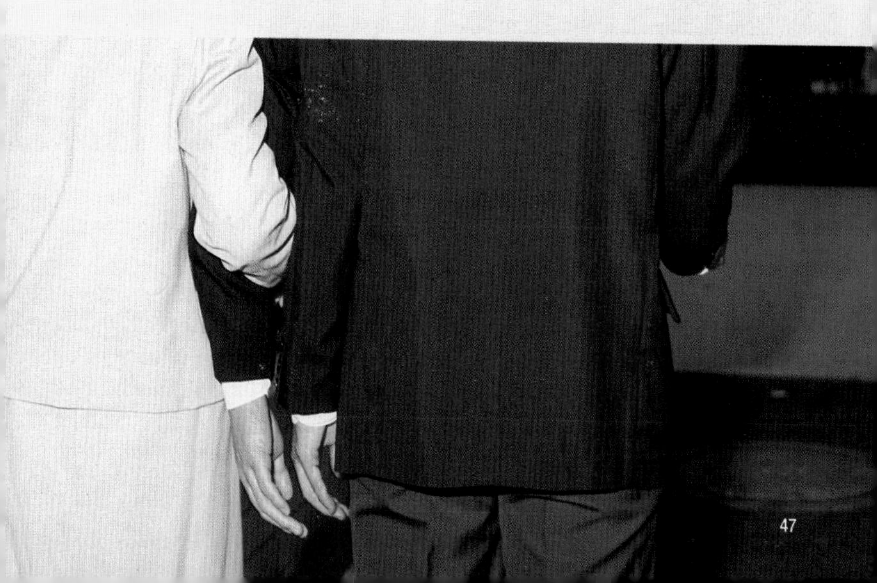

Ice Stopper

✱ *Protects the drinker's nose and lip from floating ice*

As everyone knows, ice floats. This means that the otherwise pleasurable activity of taking a drink on the rocks is fraught with the danger of ice cubes colliding with your nose and lip. With this washable, re-usable, one-size-fits-most-drinking-vessels ice-stopping net we can be freed from this anxiety at last. The Ice Stopper even protects from the most tiresome variety of face-ice collision: the last-sip sliding cubes syndrome that occurs when the base of an almost empty glass is raised above the horizontal.

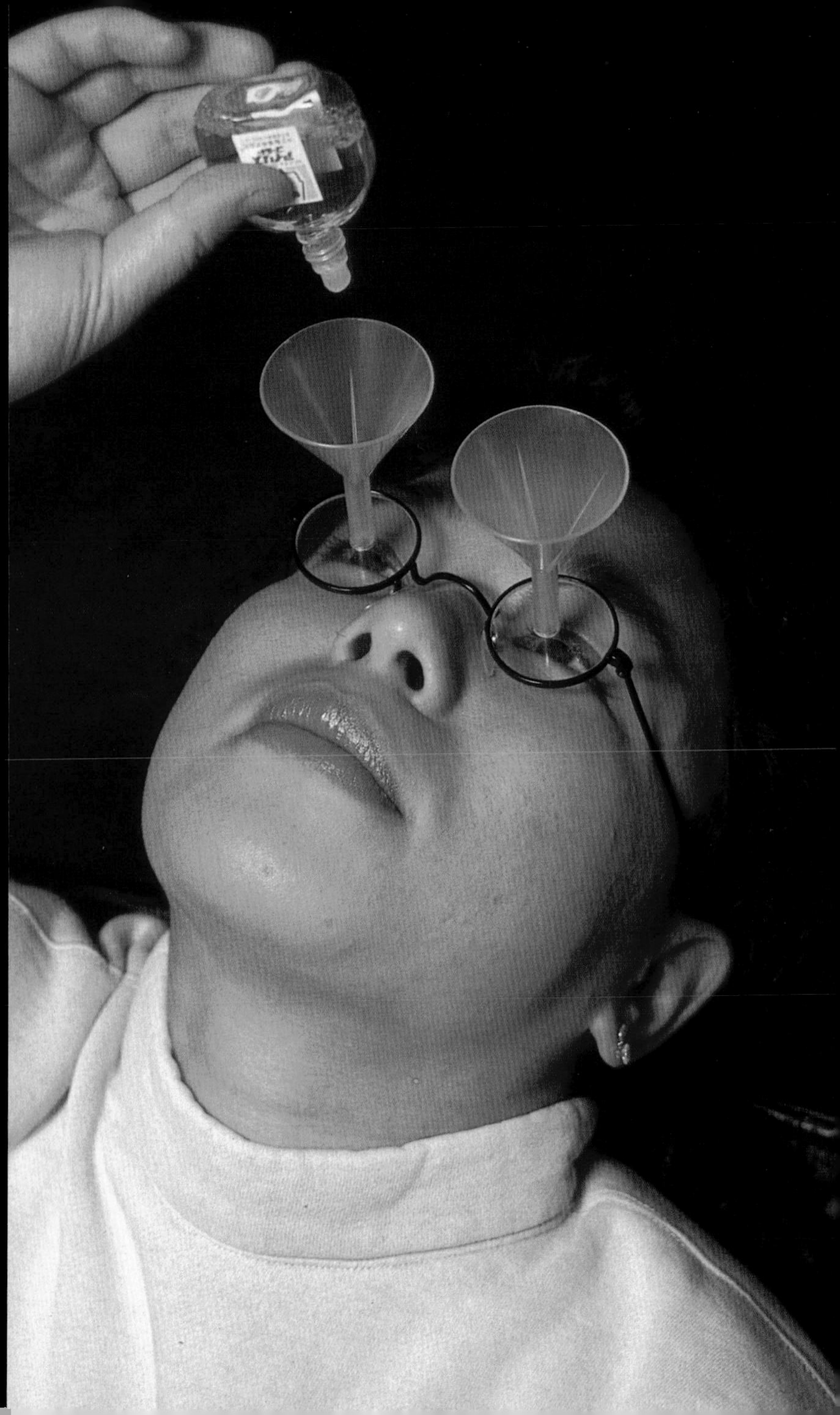

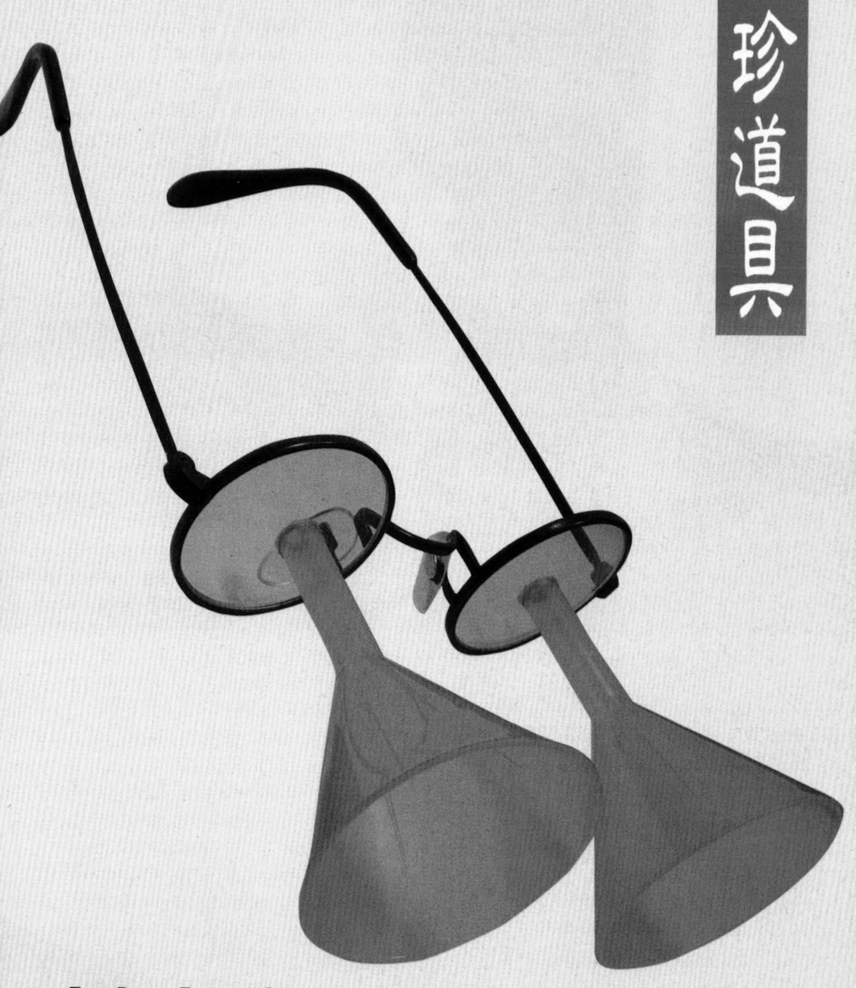

Eye Drop Funnel Glasses

✷ *For pupil-point accuracy*

Why is the application of eye-drops so problematic? You open your eyes wide, stare straight up at a bulging droplet, squeeze the bottle, and wait for gravity to do its job. Yet the next sensation you have is of the cool expensive medicine trickling down the side of your face. It's enough to make you weep.

These specially customized glasses guarantee no-more-tears eye drop application. No need for accuracy. Provided the solution makes it into the funnels, you'll score a bull's-eye every time.

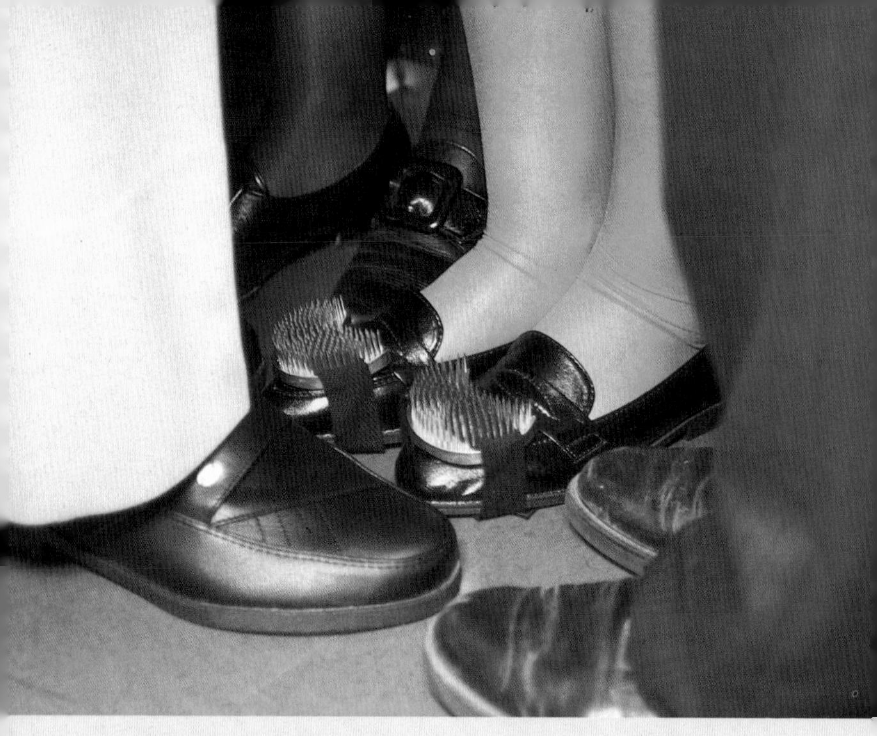

Shoe Guards

✱ *Commuter resistant protection for well-shined shoes and sensitive toes*

In the twice-daily commuter crush, it's often your feet that come off worst. Even those fortunate enough to escape a toe-crushing can expect to lose the gloss on their well-buffed shoes. Now there is a dramatic but highly effective solution to both problems. Anyone incautious enough to step on these steel-spiked toe-guards will not make the same mistake twice.

If only Elvis had had a set, one feels his blue suede shoes would have been quite safe.

Lazy Grabber

✱ *The long arm that saves your legs*

It's all very well having labour-saving devices like the TV remote control. But what do you do when they stray out of reach? You sit and suffer until you have to get up for something really important anyway.

The lazy grabber alleviates such stress, by allowing you to retrieve such items without moving from the comfort of your favourite chair. It may, however, be advisable to have two Lazy Grabbers – just in case one of them strays out of reach.

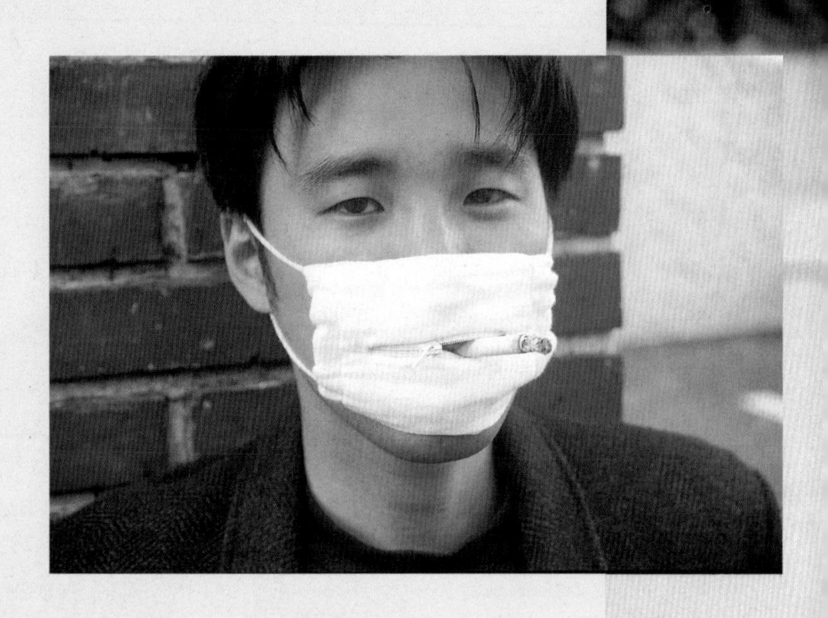

Zip-up Cold Mask

✱ ***For instant oral access***

The cold mask has long since shown its worth in keeping germs at bay. This popular accessory has, however, proved something of an inconvenience at mealtimes. The current options are either to remove the mask completely, thereby exposing yourself to foreign germs for the duration of your mealtime, or pull it down from the mouth for each sip or bite, causing uncomfortable elastic strain on the ears. As in so many areas of life, the zip fastener offers a simple solution. The Zip-up Cold Mask allows you to eat, drink, and even smoke, with minimal exposure to contaminated air.

Fresh Air Mask

✷ *The natural remedy for bad city air*

We all like to get out of the smoke for some clean country air. But for those who manage to escape less often than they'd like (practically all of us), there is another way. By bringing a living plant to your personal air space, the Fresh Air Mask ensures that Mother Nature has every opportunity to cleanse the air you breathe as only she knows how. A bonsai forest version of the Fresh Air Mask is currently in development.

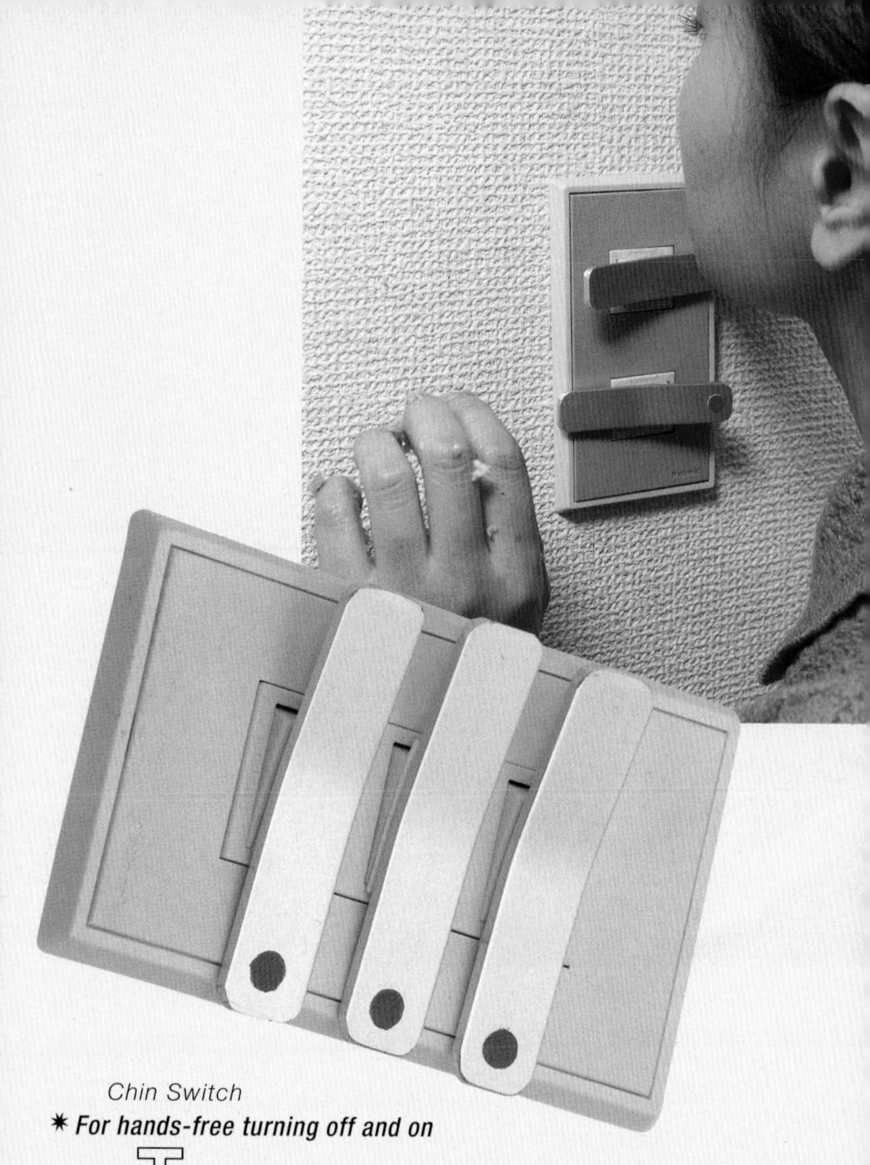

Chin Switch

✷ *For hands-free turning off and on*

There are numerous occasions when the fingers may not be the best part of the body with which to turn on a switch. Perhaps they are unavailable (for example when you are carrying a precious work of art). Or they may be wet, or covered in sticky rice or tempura batter.

The Chin Switch is specially designed to be activated by one part of you that is almost certain not to be engaged in another task: your chin. Men who happen to be shaving will find that it is also easily activated by the elbow or forehead. Or finger.

Peanut Deshelling Bag

✷ *You get the nut, the bag gets the shell*

The peanut is a great treat, and never more so than when freshly sprung from its natural casing. The aftermath of a good shelling session is, however, the tidy housekeeper's worst nightmare. Bits of peanut shell find their way into the most unlikely places.

The Peanut Deshelling Bag may sound an unlikely enough place itself, but once inside it, nothing can escape – except the meaty kernel that is the true object of desire.

There are two openings for your hands, ample storage for a feast of nuts, and a clear vinyl window that allows you to keep an eye on your work. The bag is easily emptied after a session, but be careful not to spill shell fragments through the hand holes on the way to the waste bin.

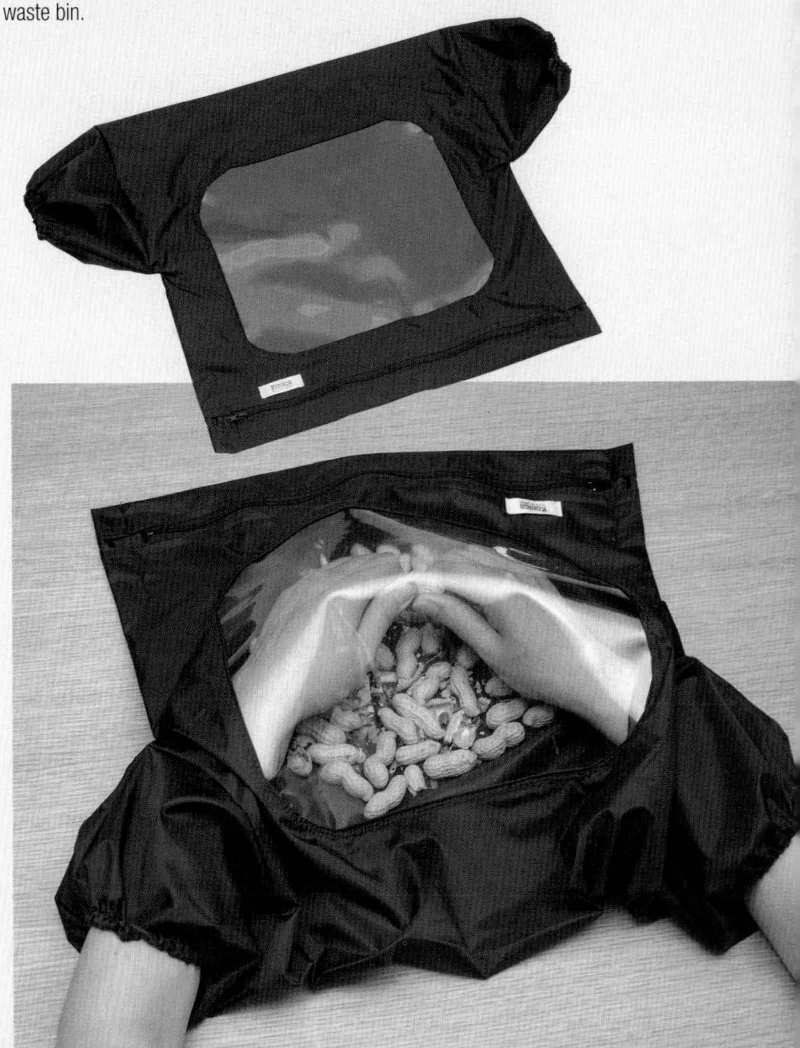

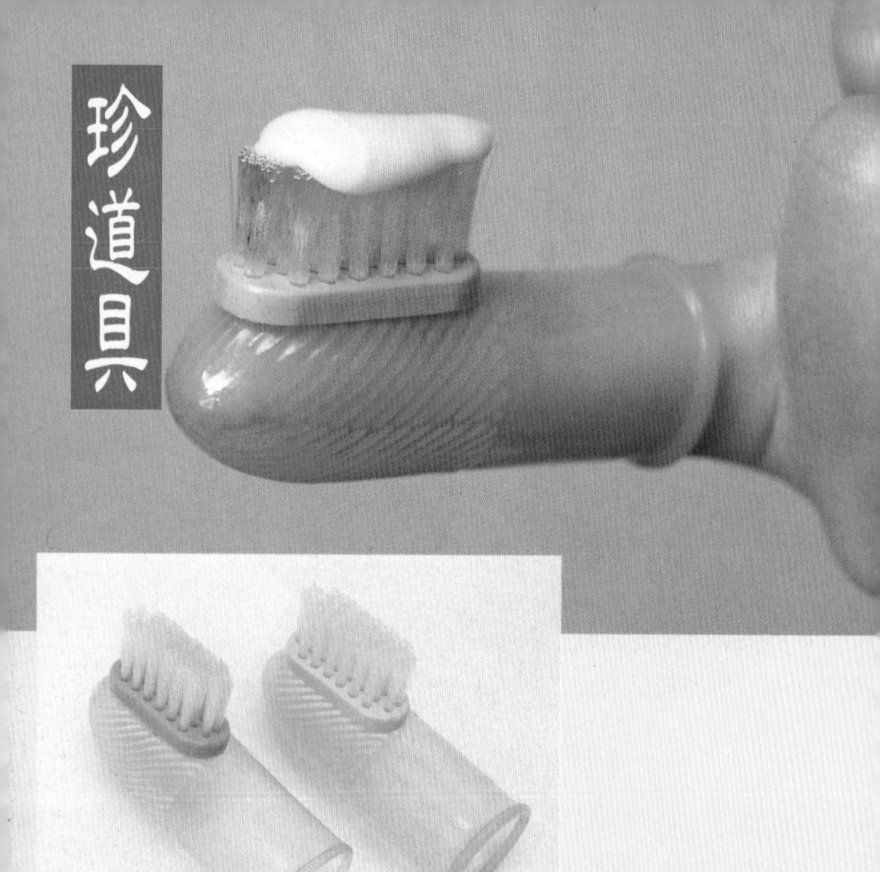

Finger-mounted Toothbrush

✷ *Oral hygiene at your fingertips*

When it comes to perfect oral hygiene, there's one thing that always gets between you and your teeth – the handle of your toothbrush. We have removed this cumbersome impediment, and made toothbrushing a more hands on, or fingers on, experience. You can vary the pressure according to the sensitivity of individual teeth, and get in close to do more delicate work.

Author's note: Rumours have reached us that the fingertip toothbrush actually exists as a marketed product. If this is true, then it is technically disqualified from being a Chindogu. However, should the product fail to catch on and be withdrawn from the market, as we might anticipate, then its status as Chindogu may be restored.

Dust and Shake

✷ ***For well-earned after-chore refreshment***

You've heard of Shake 'n' Vac, but here's an even better way to put the freshness back into tedious household chores. Before dusting, fill the Dust and Shake with three parts gin, a dash of Martini, and half a dozen cracked ice cubes. Dust vigorously, then strain your chore-rewarding dry Martini into a glass, and put your feet up. Then get up, refill, and start on the next room. Your household chores will soon begin to seem like one long party.

Konica

Umbrella Tripod

✷ *The photographer's friend come rain or shine*

This essential aid for amateur photographers keeps the rain off your camera (in much the same way as a conventional umbrella), but when the sun comes out converts instantly to a stable tripod (strictly speaking an octopod, since the 'legs' on which it stands are the eight spokes of the umbrella's canopy). Ideal for those self-timer group shots, the Umbrella Tripod makes it even easier to smile for the camera.

Portable Stoplight

✷ *Puts traffic control in the hands of those who need it most*

Children do not always remember to apply the traffic safety code, and this is an increasing source of anxiety for parents. But now they can relax again – provided they have equipped their children with the Portable Stoplight.

The child in charge just switches it on and, regardless of whether he or she remembers to look both ways, a bright red light beams in both directions, halting the oncoming traffic. Despite its authentic design, the portable stoplight only shines red. The danger of carnage resulting from an accidental green light is thus cleverly avoided.

Magnetic Slippers

✱ Helps a pair stay a pair

How much time do we waste each year looking for a lost slipper? And how much distress is caused to fastidious members of society by the sight of two slippers not properly aligned, maybe even several feet apart? The deft use of a few small magnets has righted these wrongs. The slipper wearer simply clicks his or her heels together before taking off the slippers, and the pair remains in perfect togetherness until they are next put on. The slippers have the bonus function of picking up dropped pins and other small metallic objects as you walk around the house.

Vase Taps

✷ *Changes flower water quickly and cleanly*

Cut flowers rarely last as long as you hope, and one reason for this is certainly the tedious labour associated with changing the water. First you have to take out the flowers (which may have slimy stems), then throw away the old water, then fill the vase with fresh water, and finally replace the flowers. This is no doubt as stressful to the flowers as it is to you. The vase tap cuts these four difficult stages into two easy ones: empty, and refill.

It is particularly worthwhile to have a Vase Tap fitted to a priceless antique vase, as it greatly reduces the risk of breakage during a water change.

Table Groomer's Hand

✷ *Makes discretion the better part of oral hygiene*

A common problem encountered when dining in company is the small food particle lodged between the teeth. The traditional approach to this dilemma of social form is to attempt to remove the debris with the fingers of one hand, while masking your action with the other hand. This has never been satisfactory, as it leaves no hands free for the kind of social behaviour, principally eating and gesticulation, that would best disguise the unpleasant events.

So we bring you the Table Groomer's Hand, a moulded silicon facsimile which can be used to shield your clandestine cleaning activities. It is held up by a knob that fits between the fingers, yet leaves those same fingers freedom of movement to pick at the offending morsel. Meanwhile your third (real) hand is free for emphasis, coffee drinking, or playing games, whatever the occasion demands.

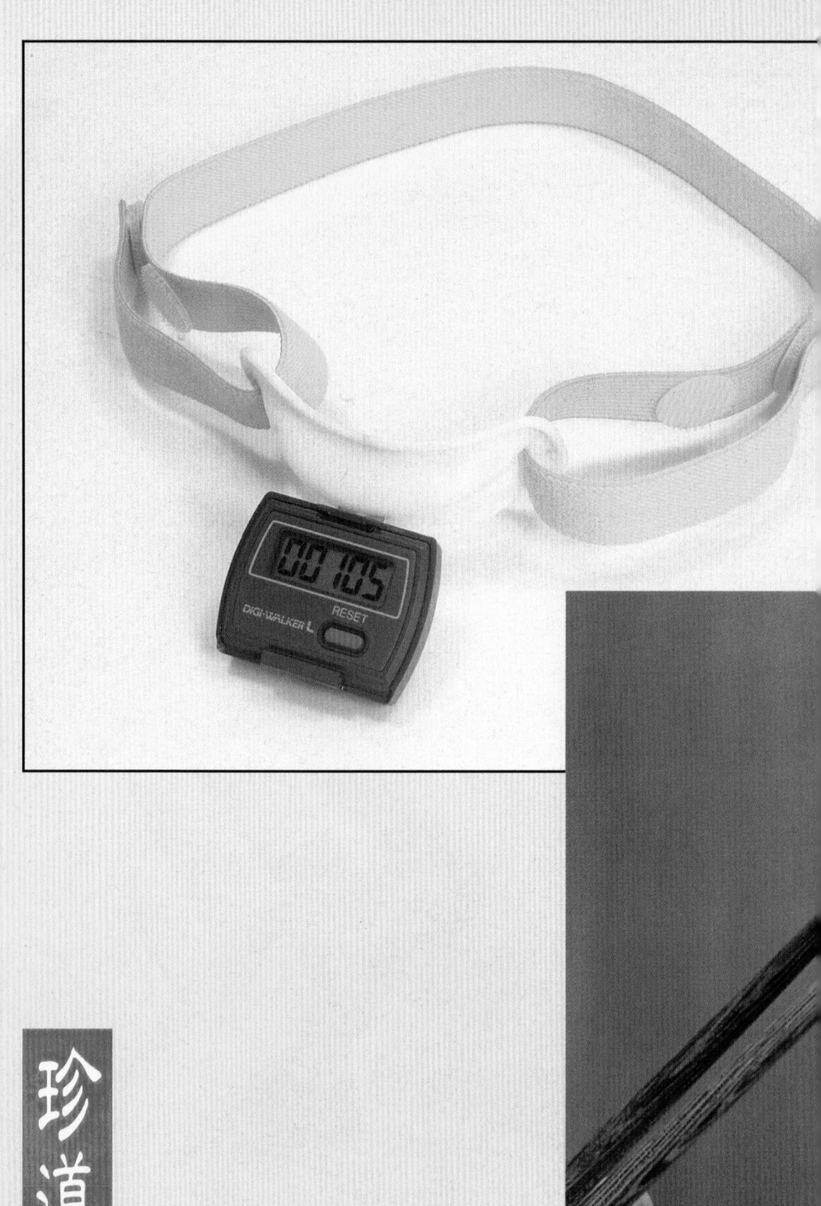
00105
RESET
DIGI-WALKER L

珍道具

Automatic Chew Counter

✷ *No more too few chews*

Today's food presents considerably less resistance to the teeth than that of our ancestors. Even the modern hamburger can practically be sucked through a straw. The consequence of this in the general population is lazy eating, ailing jaw strength and, when we do encounter foods that could do with a chew or two, indigestion. We know that a regime of proper chewing would rectify such problems – and an estimated 2,000 chews per meal is what the experts recommend. But who can be bothered to count?

The Automatic Chew Counter, that's who. Just strap it on at the beginning of a meal, and check the counter at the end. If you've fallen short of the requisite 2,000, then strap it back on and eat some more. Those on a diet are recommended to make up the deficit by vigorous 'air-chewing'.

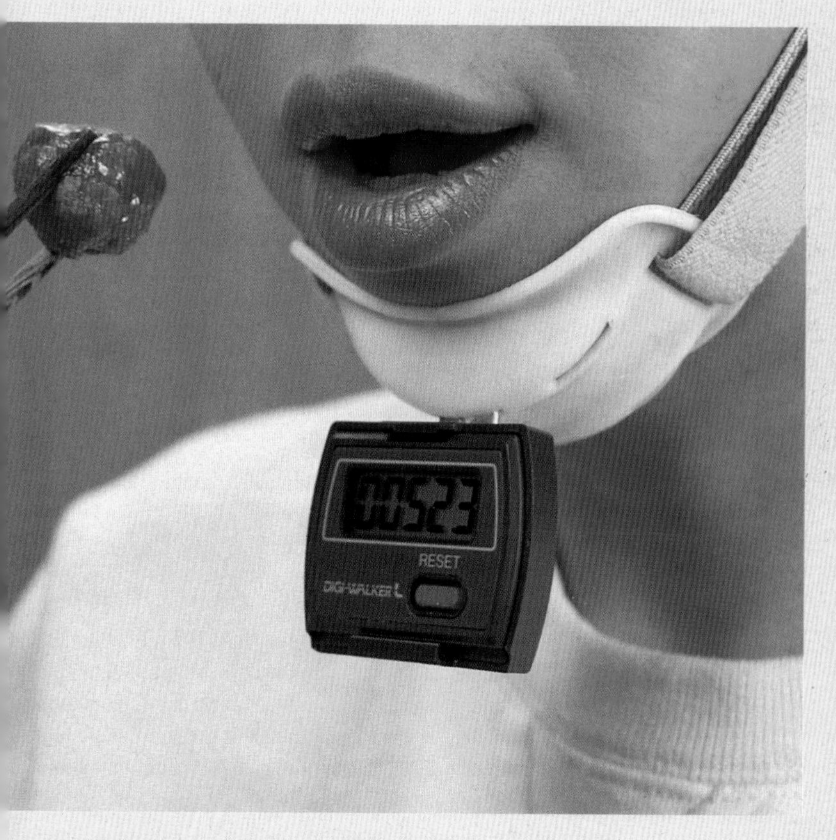

Solar-powered Lighter

✷ *Light up while the sun shines*

A stroll in the park, a smoke on a bench – it's the perfect way to relax in your lunch hour. But stress and frustration come flooding back when you realize that your lighter is out of gas, and you've left your matches in the office. For the proud owner of the Solar-powered Lighter, all is not lost – provided the sun is shining. (In Britain, exclusive use of the Solar-powered Lighter is particularly recommended for anyone wishing to cut down their smoking habit dramatically.)

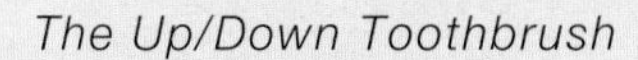

The Up/Down Toothbrush

✷ *Cuts brushing time in half*

With bristles running both ways, the logic of this labour-saving toothbrush is clear – only half the strokes are required to clean both top and bottom sets. The time-saving benefits are less apparent when the front and back as well as the ends of the teeth are taken into account. But at least the inside of your lips and the front of your tongue will also get a good clean.

Hay Fever Hat

✷ *The all day tissue dispenser*

Having hay fever is bad enough, but running out of hankies turns misfortune to misery. So, don't run out! The Hay Fever Hat supports a large loo roll, enough to cope with heavy blowing and incessant sneezing from dawn till dusk.

Portable Commuter Seat

Reserve a seat – take it with you

We all like to get a seat on a crowded train. But some of us don't want to fight for it (and some of us who do want to fight for it always seem to lose). As so often in life, the solution is to bring your own. The Portable Commuter Seat fits snuggly in the slight space between the thighs of more conventionally seated passengers. In fact the slighter the space the better the ride, as a tight fit means less wobbling will occur when the train goes round the bend.

Umbrella Shoe Savers

✱ ***Extends the life of expensive footwear***

The conventional umbrella offers proportionately decreasing protection from precipitation for the lower parts of the body. The latest in the Chindogu range of improved umbrella technology goes right where you need it most – on the long-suffering shoe. The mini-canopies have a full thirty-centimetre diameter guaranteed to keep the rain off the full front of foot area. Take off the umbrellas when you get where you're going, and your bone dry shoes look as good as they did when you put them on.

Note: the generous width of the umbrellas means that care must be taken when walking not to bring the feet too close together.

Vanity Camera

✷ Allows last minute adjustments for the camera-shy

We all like to look our best for the camera. But who would trust a photographer to know when we look good? With its built-in mirror, the vanity camera gives the subject ultimate control. We can fine tune that smile, flatten stray hairs, and choose our best side, right up to the moment when the camera goes click.

Self-cooling Hat

✷ *Helps hot heads stay cool*

If you can't get ahead, get a hat. But what can you get if your hat's too hot? The Self-cooling Hat, of course. The attached fan pumps out hot stale air from inside the hat, and draws in fresh cool air from outside the hat. It's powered by a self-charging solar battery, which stays on when the sun goes in for a while, or the wearer walks in the shade, but finally gives out only when the cool air of the evening has arrived.

Flotation Bag for Water Reading

✱ Keeps the reader's head above water

Having already solved the problem of reading in the bath (see page 16), Chindogu has taken on an even bigger challenge: the problem of reading while wading across large bodies of water. But why should anyone want to read while wading across large bodies of water? Because now they can!

With rubber boots attached to rubber trousers and an inflatable rubber ring, we've formed a waist-high underwater suit. In work or leisure clothes, you can get some relaxing float reading in at any time of day, and stay dry as a bone. And of course you don't have to wait till you encounter a body of water that needs to be crossed – you can do it for fun!

The Driver-drier

✷ *Improves the swing, dries the laundry*

Another two-in-one Chindogu labour saver, the Driver-drier not only does a fine practical job, it also goes a long way to restoring matrimonial harmony in households where golf widows have reached a point of despair. The golfing husband gets to practise his swing, while his wife has the satisfaction of finally seeing her partner contribute to the mountain of household chores.

Novice swingers should be careful not to swing those pristine whites too close to the grass, or the effect on matrimonial harmony may be swiftly reversed.

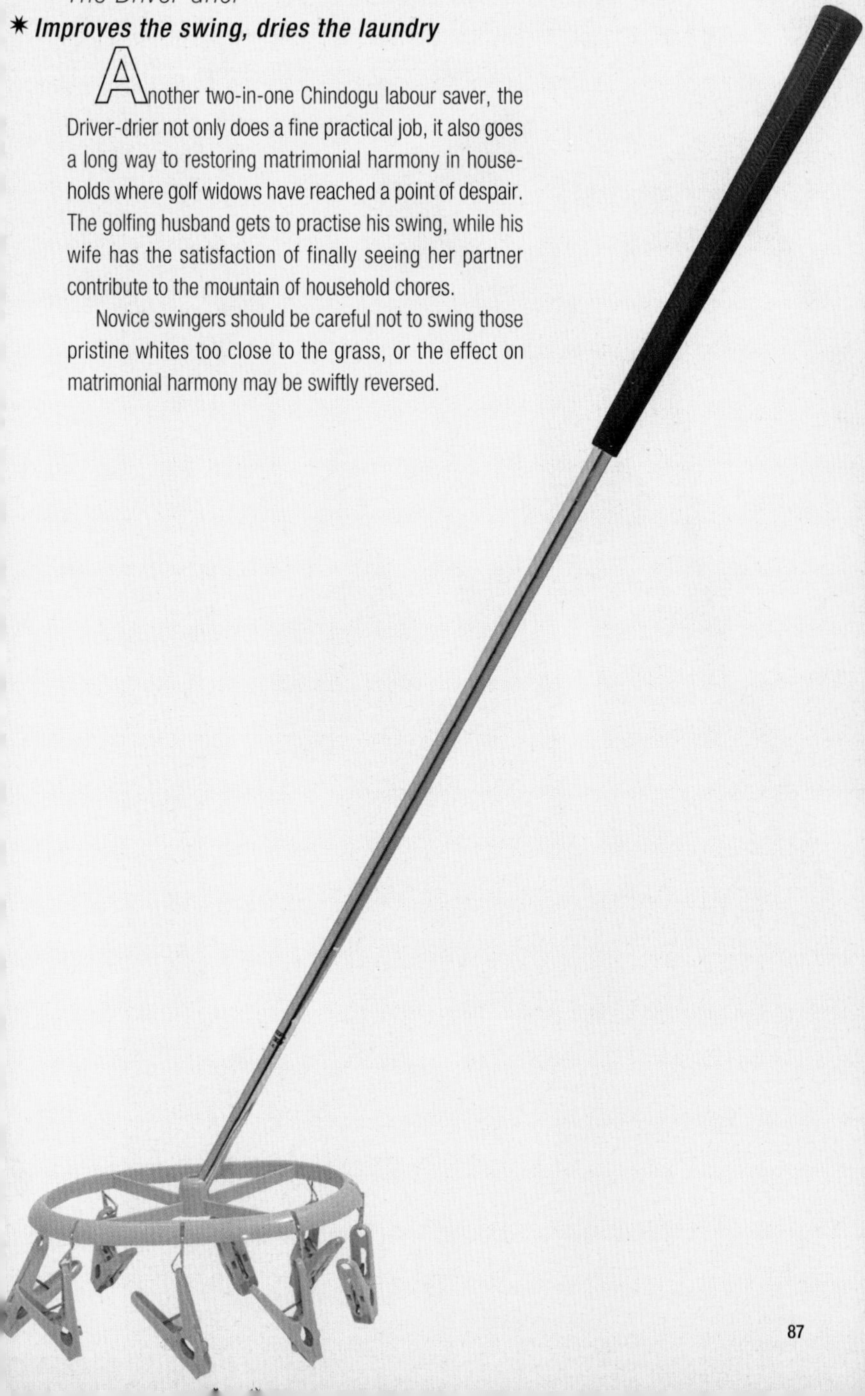

DIAMOND HOTEL
DDI

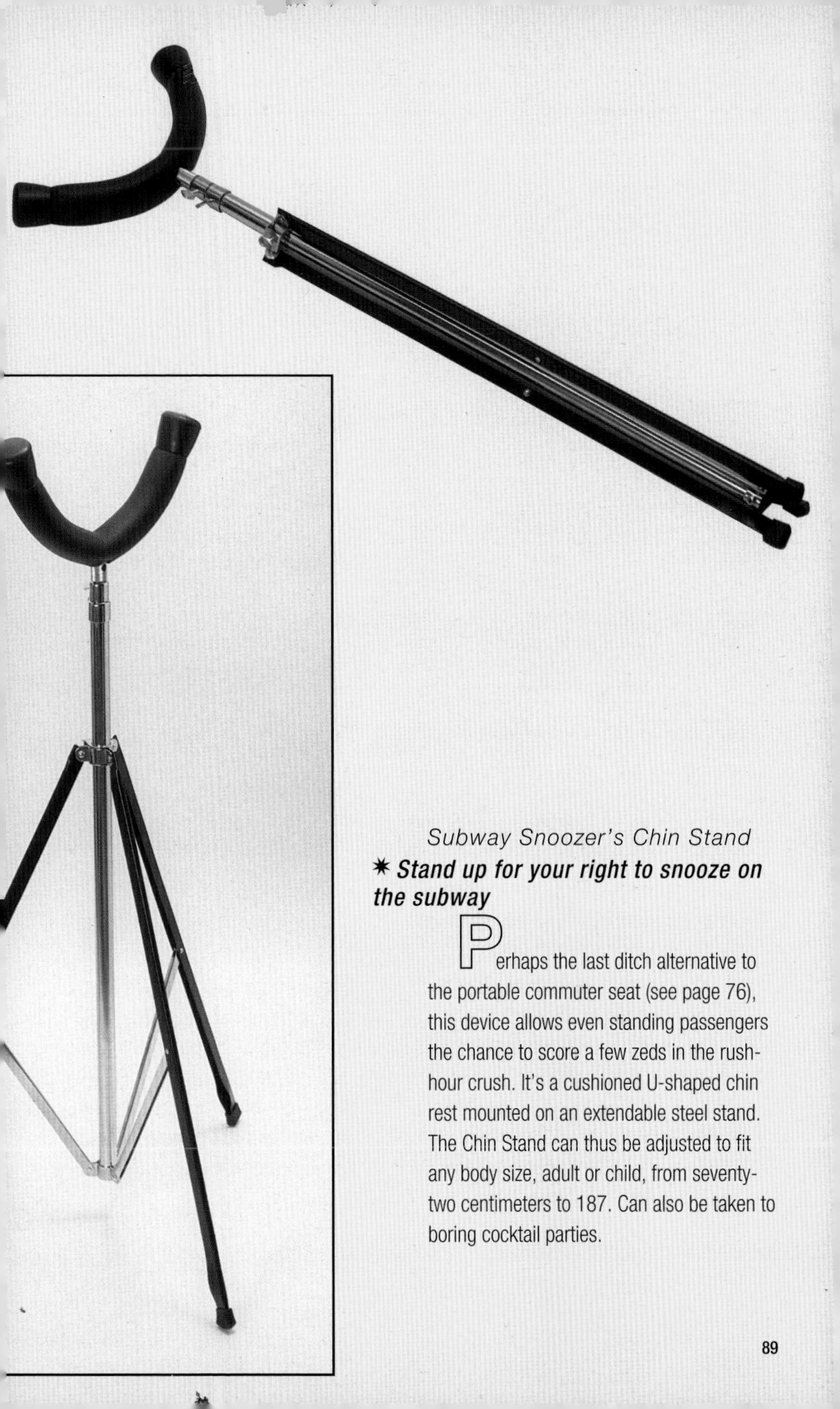

Subway Snoozer's Chin Stand

✷ Stand up for your right to snooze on the subway

Perhaps the last ditch alternative to the portable commuter seat (see page 76), this device allows even standing passengers the chance to score a few zeds in the rush-hour crush. It's a cushioned U-shaped chin rest mounted on an extendable steel stand. The Chin Stand can thus be adjusted to fit any body size, adult or child, from seventy-two centimeters to 187. Can also be taken to boring cocktail parties.

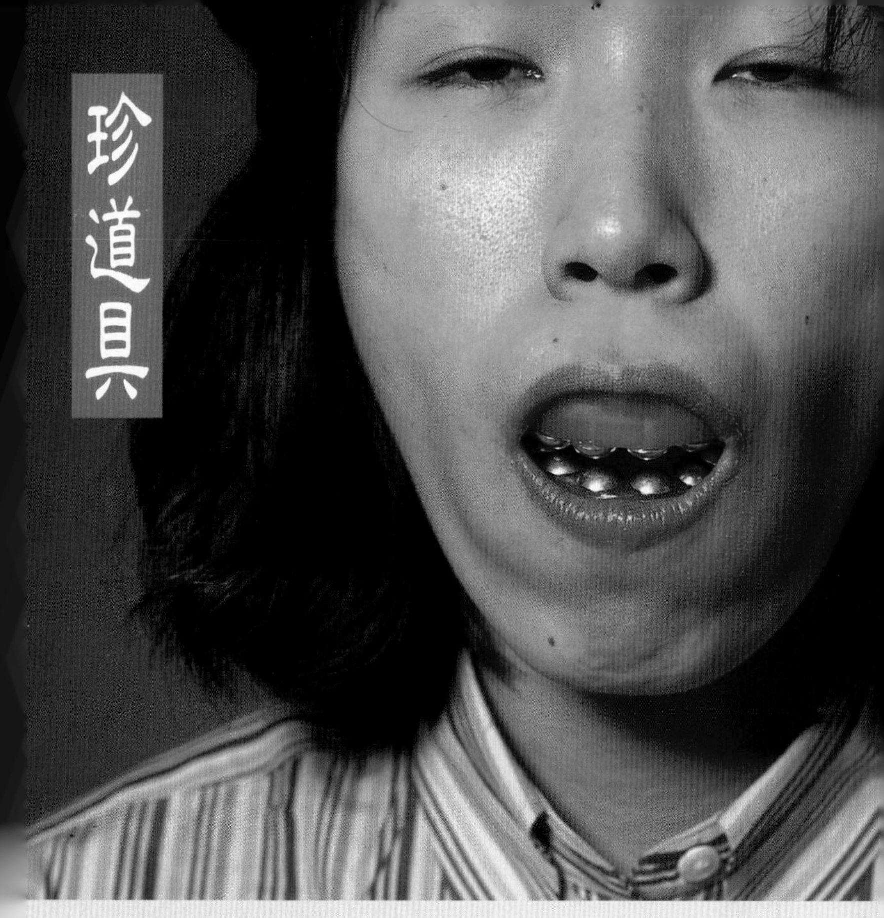

Detachable Tooth Covers

✷ *Saves brushing after every meal*

After putting so much effort into a rigorous morning routine of dental hygiene, it can be distressing to find all that hard work undone by the midday meal. Detachable Tooth Covers put an end to contamination by food of pristine teeth. The gumshield-style silicon plates fit comfortably in the mouth and are easily attached and removed. Each plate is fitted with seven hygienic and hardwearing stainless steel artificial molars which allow you to crunch and chew without food coming in contact with your natural teeth. For those who like to keep their Detachable Tooth Covers in pristine condition, we are now developing Detachable Tooth Cover Covers.

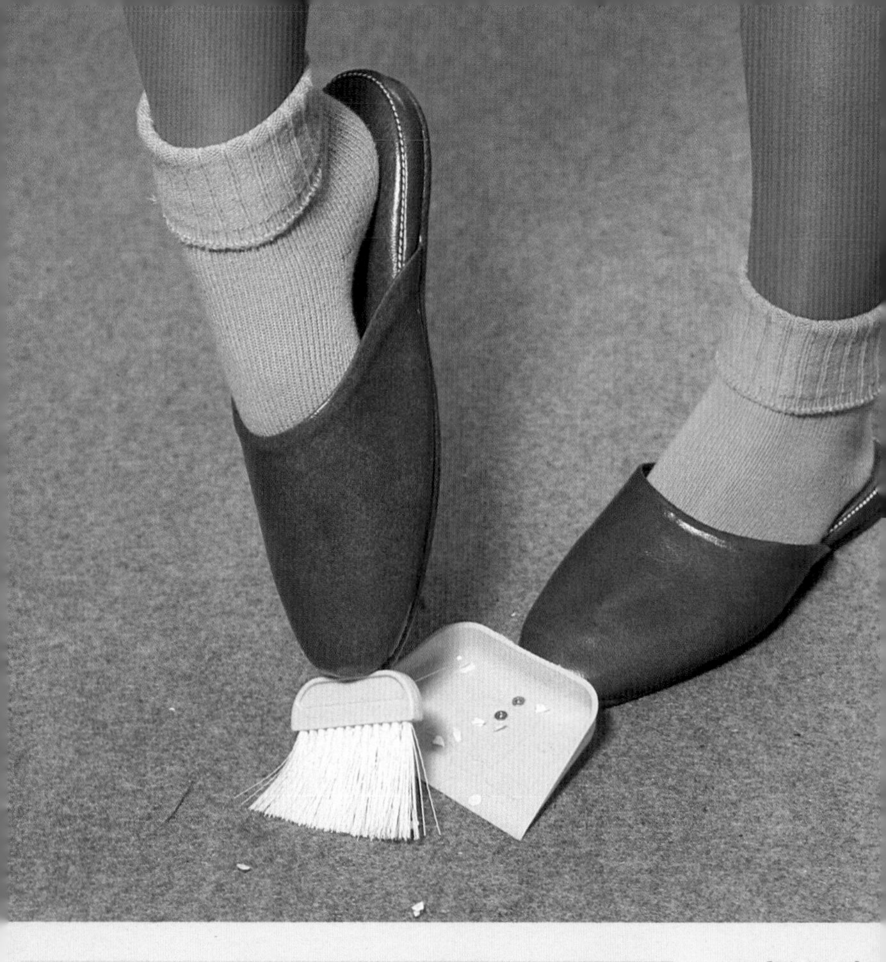

Clean-up Slippers

✷ *Let nifty footwork do the housework*

A little piece of fluff or a scrap of paper on the floor hardly justifies wheeling out the vacuum cleaner, or even rushing to the cupboard for the dustpan and brush. But it is nonetheless unsightly, and a source of irritation to those who like to keep the house just so. Clean-up Slippers have a left toe-mounted mini dustpan and a right toe-mounted mini brush (reversable for the left-footed), which allow the wearer to deal with even the smallest blemish as it is encountered. This way a stroll around the house practically does the cleaning for you.

Note that any matter gathered in the dustpan should at once be disposed of in the rubbish bin, as by walking around after sweeping up, items of dirt can be redistributed inadvertently elsewhere in the house.

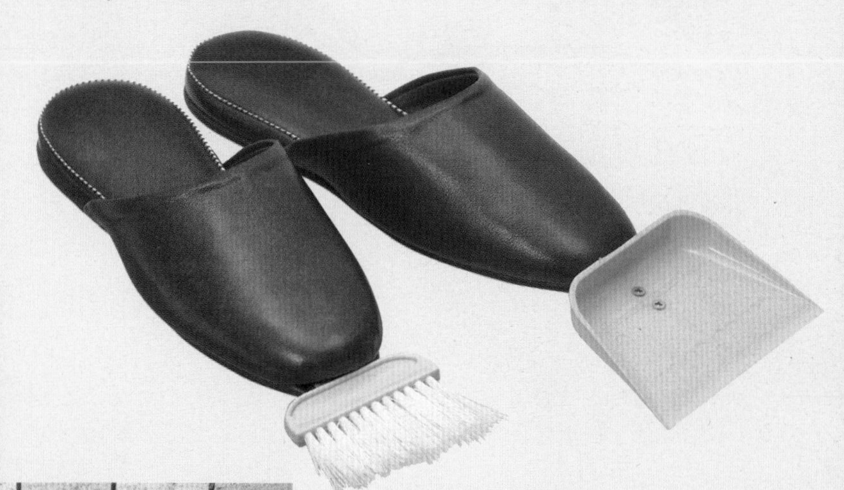

Self-stirring Frying Pan

✷ *Master of the stir-fry*

Up until now, the stir-fry gourmet has always required two instruments to pursue his or her culinary art: a frying pan, and a stirring utensil. The TV chefs make it look so easy, but in practice it requires a level of co-ordination beyond the ability of most home cooks. And that's why we built the Self-stirring Frying Pan. Its non-stick surface and rotating double spatula keeps your ingredients constantly on the move – all you have to do is turn the handle.

Cockroach Swatting Slippers

✷ *For longer range and greater accuracy*

Research has shown that the slipper is already the preferred instrument for indoor cockroach swatting, with up to eighty per cent of householders preferring indoor footwear to rolled up newspapers or paperback books. But why not improve your hit rate by investing in a pair of special Swatting Slippers? With telescoping handles you can operate from longer range with greater accuracy – no more desperate throwing of slippers as the scuttler disappears behind a sofa. Ideal for those who can't, or won't, get too close to their prey.

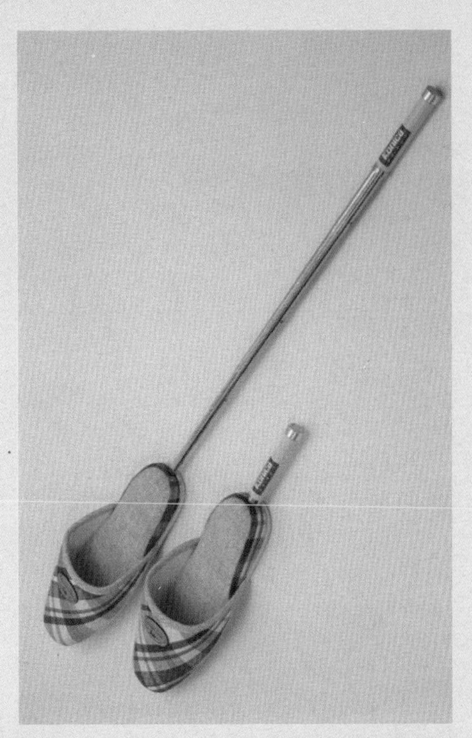

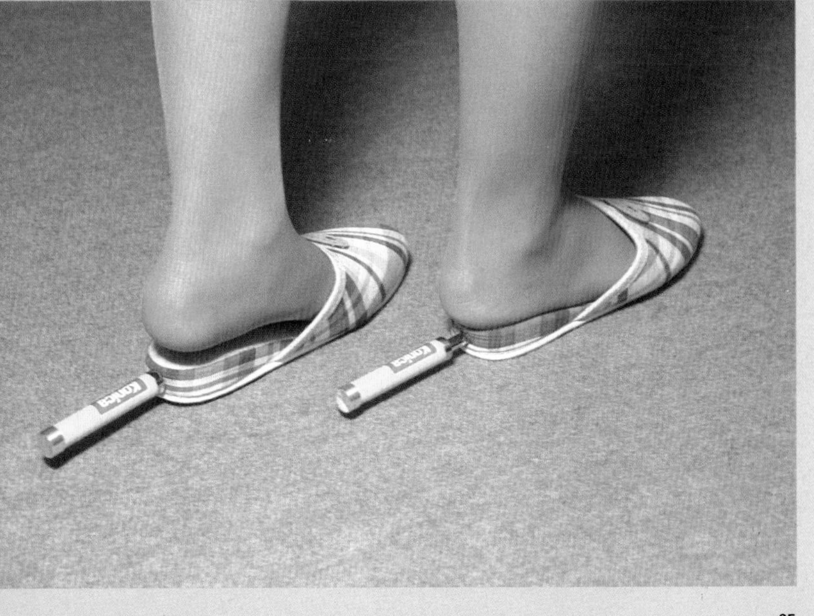
Konica
Konica

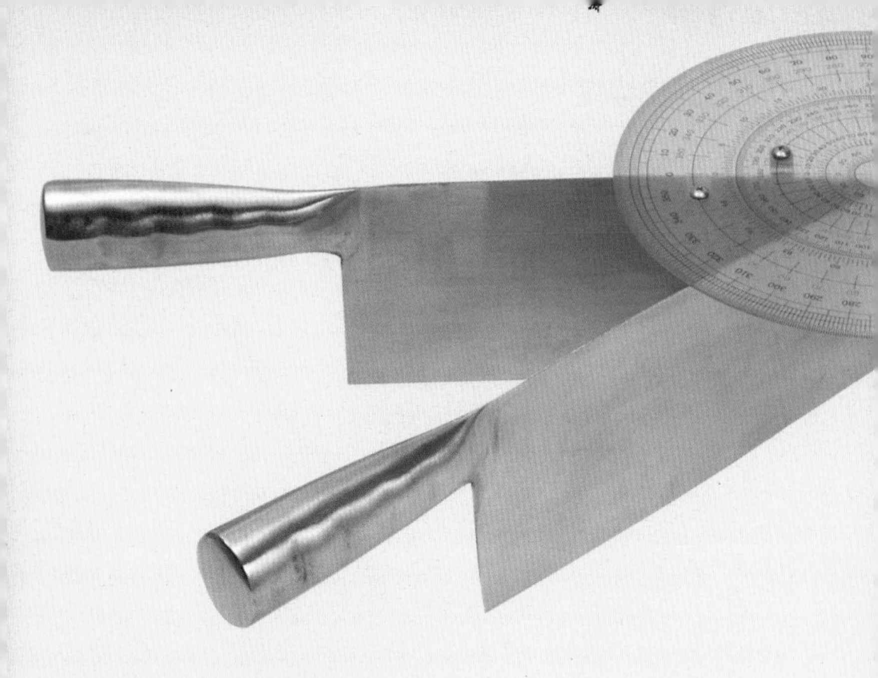

Perfect Cake Cutter

✷ *For fair shares all round*

Even the most fair-minded of hosts can come unstuck when dividing a round cake between a number of greedy guests. Dividing into two, four or eight is moderately easy to achieve by eye, six less so, and any odd number is simply hopeless. Someone's sure to feel short changed, and if kids are involved it's a no-win situation – unless you can persuade each of them that their bit is biggest.

The Perfect Cake Cutter transforms cake division into the happy family sharing situation it was always meant to be. This two-bladed device is attached to a 360º protractor, with one blade fixed on zero degrees and the other adjustable to create the required angle – 120º for a three-way division, 72º for five, and 51.4º for seven. Total fairness, of course, depends on being able to locate the exact centre of the cake, for which we suggest you use a compass and dividers.

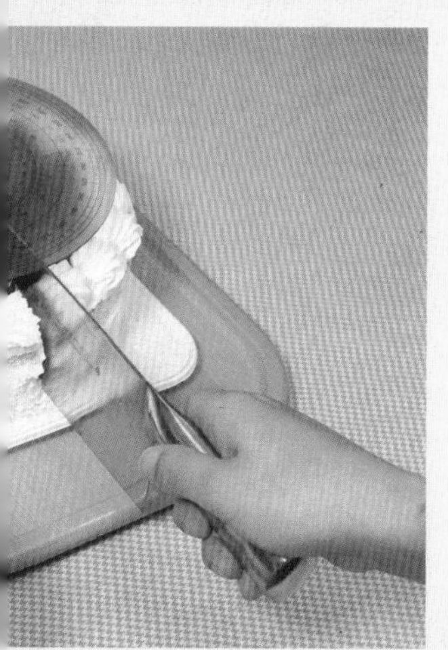

Camera Umbrella

✷ *For rain-safe picture taking*

Wet weather might be threatening to spoil your holiday, but don't let it spoil your holiday snaps. This thirty centimetre dedicated umbrella fits into and over your camera, so that even if you get rained on, your camera doesn't. A second feature of this Chindogu is a special attachment that allows you to mount the camera on top of your own umbrella, so that both camera and camera operator remain dry. The camera thus mounted can even be operated by remote – though at the small price of rendering you unable to look through the viewfinder.

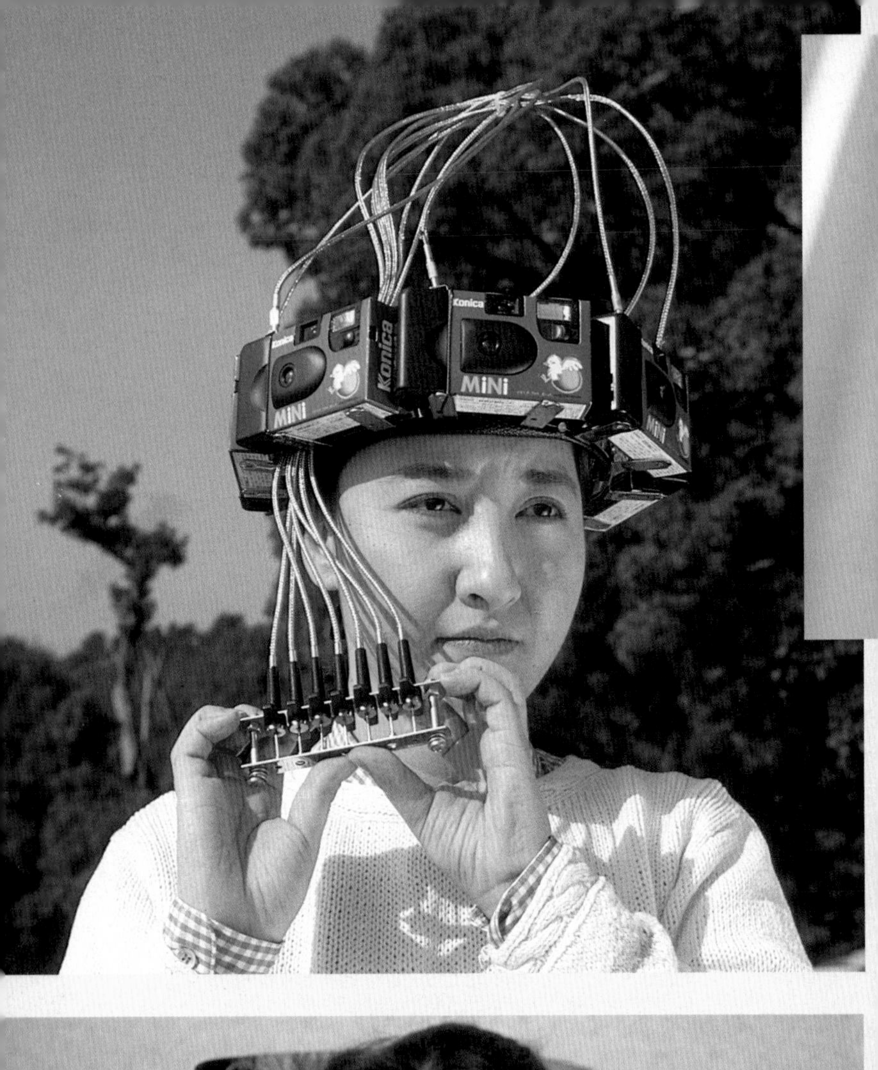
Konica
MiNi
Konica
MiNi

360° Panorama Camera

✱ *For the all round view*

A popular pastime amongst holiday photographers these days is the construction of the panoramic view. Conventionally this is achieved by taking one picture of the view, turning a little bit, taking another, until the complete view has been covered. It is, at best, an optimistic and inaccurate procedure which often yields patchy results. The Panorama Camera puts panoramic photography on an altogether more dependable footing. Seven outwardly pointing cameras are perfectly angled to ensure that all the pictures will fit together neatly. They are also fired simultaneously, a highly desirable feature which eliminates the embarrassment of having, for example, the same bird flying through two different sections of the panorama.

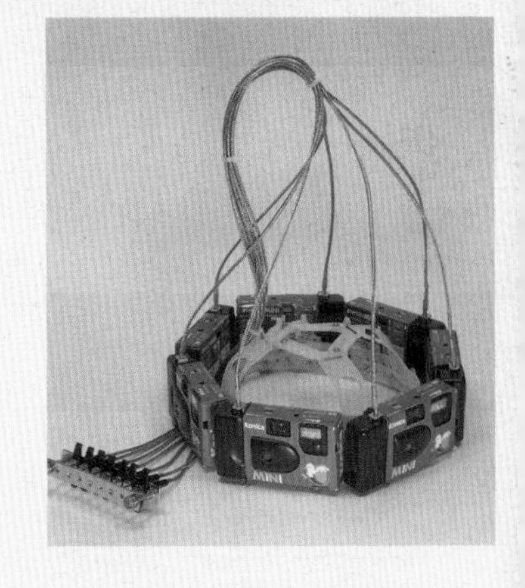

For maximum enjoyment of your panorama, tape the seven pictures together with the images facing inwards and place your head in the centre of the loop. Then revolve the loop slowly around your head.

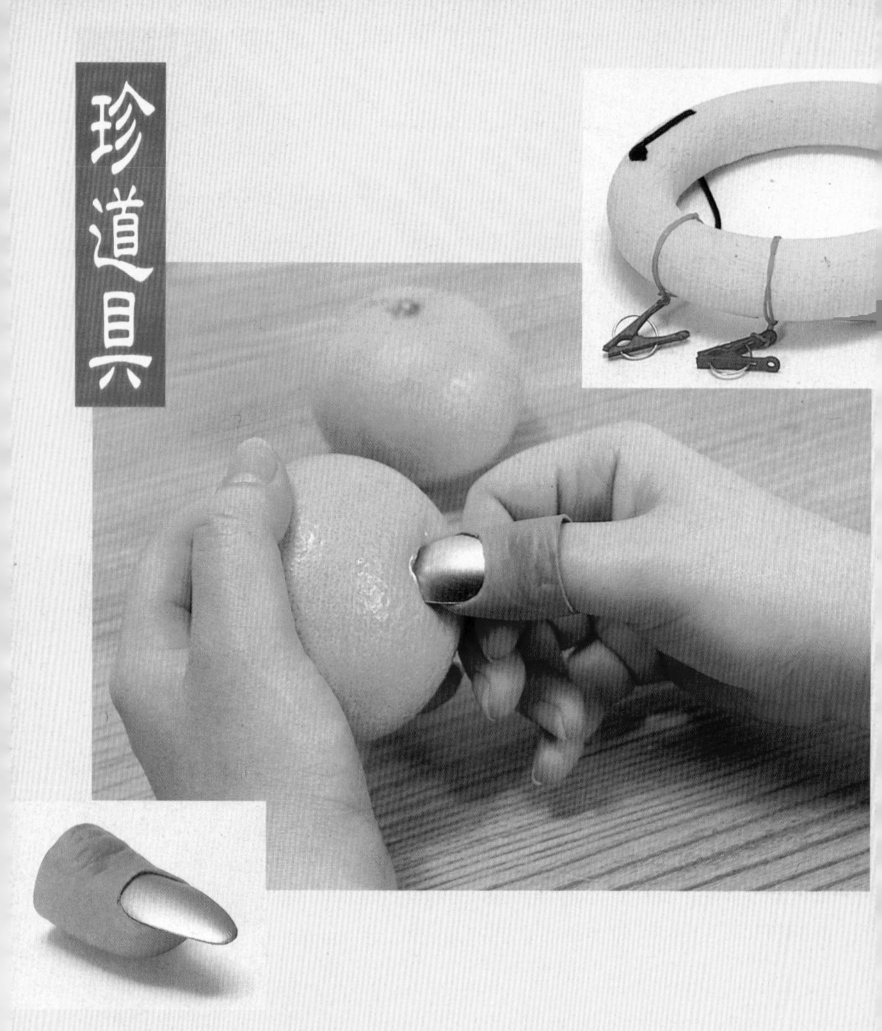

Orange Peel Thumbnails

✷ *Restores comfort and dignity to orange eating by hand*

The close-to-nature way to eat an orange is to break the skin with your thumbnail and use the same thumb to lever off strips of peel and pith till the fleshy segments are revealed. But with orange stains on your thumb, debris stuck behind your nail, and juice stinging any minor abrasions, it's a high price to pay for oranges *au naturel*. The Orange Peel Thumbnail is a robust rubber artificial thumb top with stainless steel nail. It fits over your natural thumb, allowing you to dig in to an orange with enthusiasm but without contamination. Nail-biters will also find it useful for opening penknives and scratching backs.

Wide Awake Eyeopener

✷ *Keeps tired eyes open*

When exams are looming, it often seems there aren't enough hours in the day to cram in all that vital information. So here's a way to turn sleeping time into waking and working time, without the use of caffeine or other stay-awake pharmaceuticals. Regular clothes pegs grasp the eyelids, which are held open by attachment to a head-fastener made from a rubber ring worn on the top of the head.

Those who have tried this Chindogu have complained of considerable pain – which suggests it is doing its job even better than we first thought.

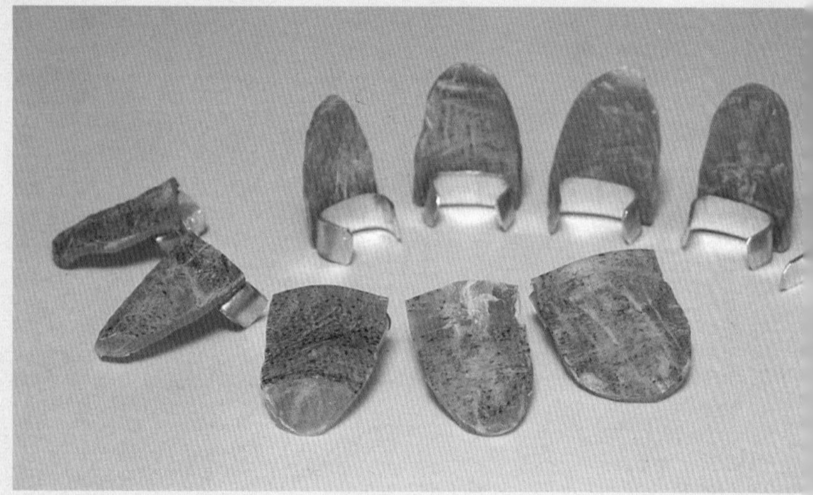

No Bite Nails

✷ *Makes not biting nails more fun*

Most treatments for nail biting depend on what psychologists call negative reinforcement. For example, nasty-tasting nail polish is supposed to provide a strong disincentive for nail-nibbling. We prefer positive reinforcement: you can bite these stick-on nails to your heart's content. Made of seaweed, they're nutritious, delicious, and surprisingly similar in texture to real human nails. A satisfying chew which leaves your own nails to grow long and strong.

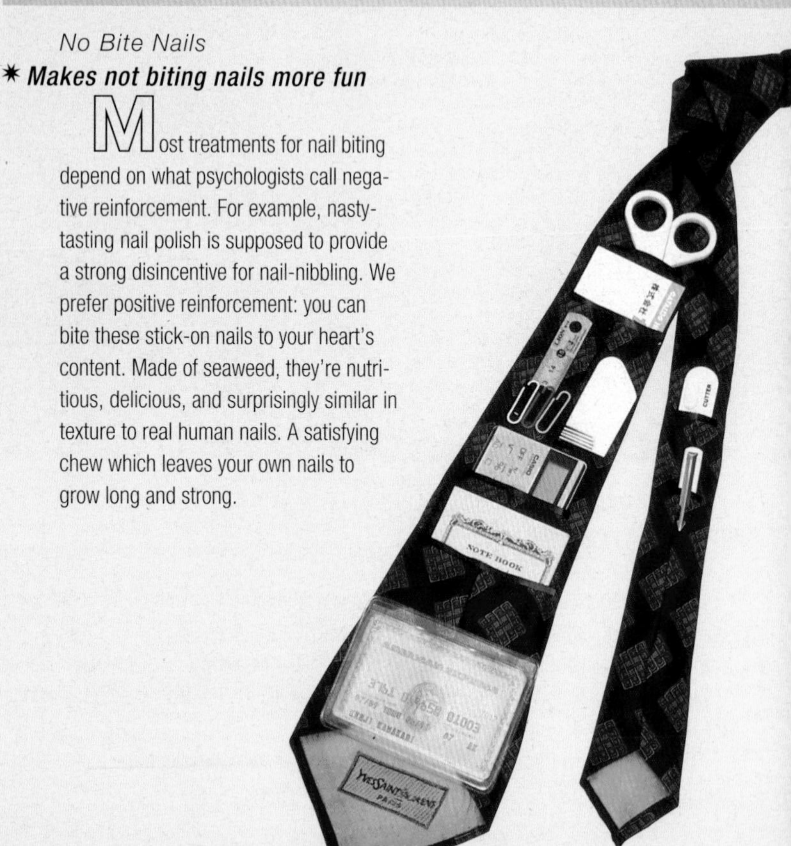

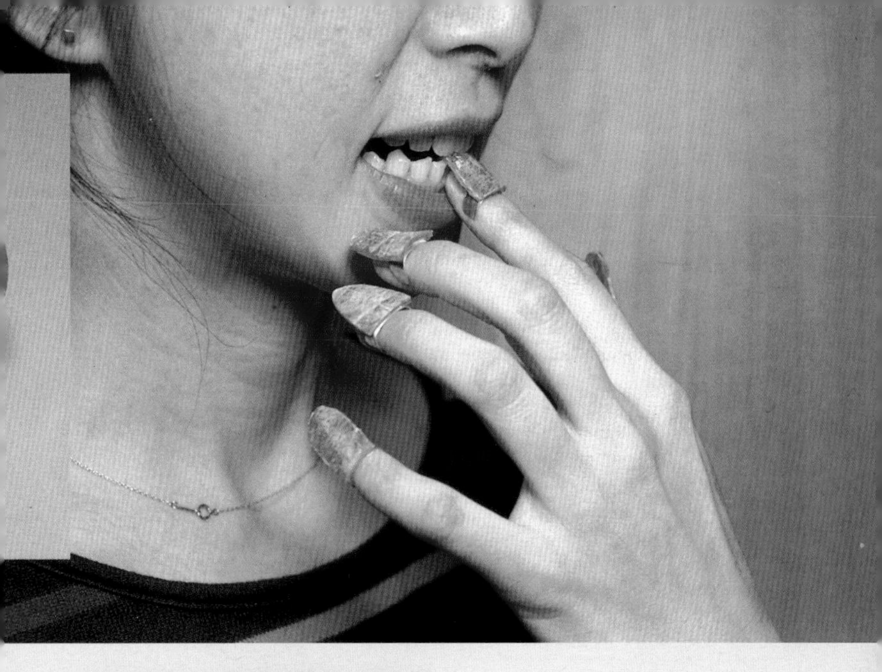

Portable Office Tie

✷ The busy executive wins by a neck

It seems somehow appropriate that that vital sartorial accessory of the businessman's uniform, the neck tie, should be put to practical use. For the real 'power tie' is not the one with the expensive designer label on it, but the one which allows you to keep ahead of business colleagues when away from your desk. The pockets and compartments of the Portable Office Tie will carry pens, business cards, calculators and notebooks – or can be customised to suit your particular line of business. It is recommended that the tie is not tied or untied until office stationery has been removed.

Earplug Earrings

✷ *Instant protecton from deafening discos*

This stylish jewelry has a well-disguised function. The attractive green pendants that hang from each hoop are in fact fully functioning rubber earplugs. Always ready, just where you need them, the earplug pendants are easily inserted without drawing attention to yourself – you can pretend you are simply adjusting your earrings. Thus you can block out the beat of blaring discotheques without being thought unhip, and turn off the sound of bossy boyfriends or too talkative teachers without being thought rude. And if anybody tells you your earrings look stupid . . . well, you don't have to listen.

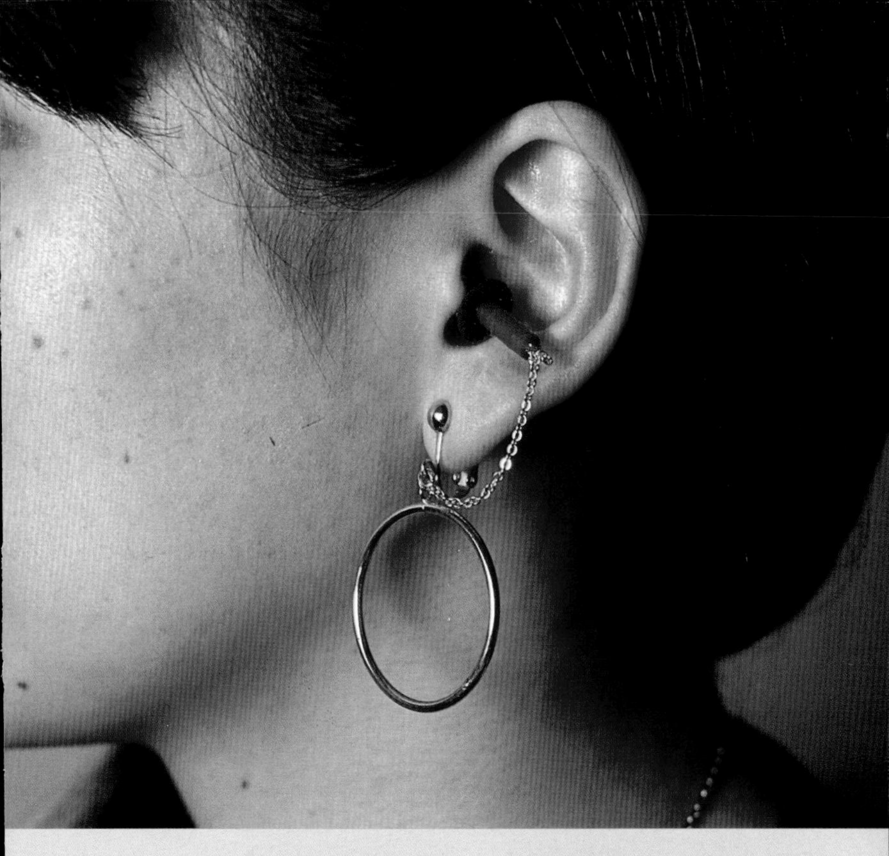

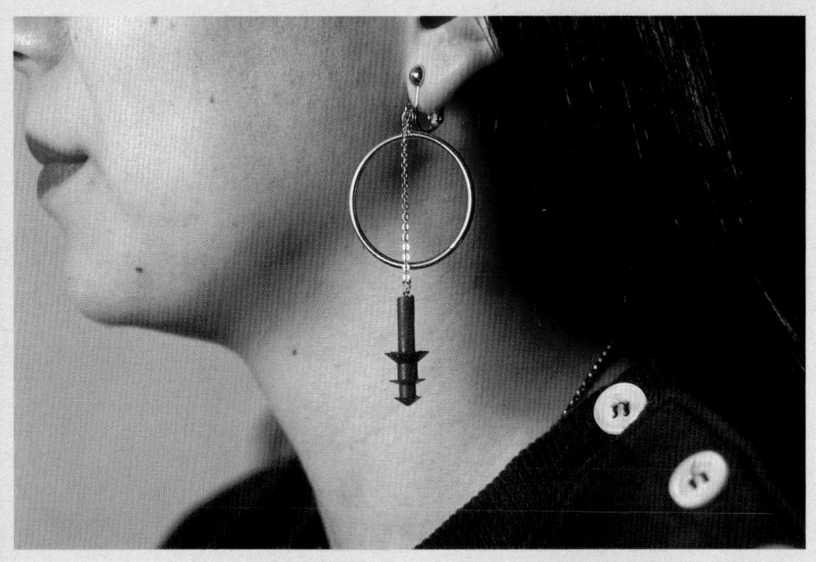

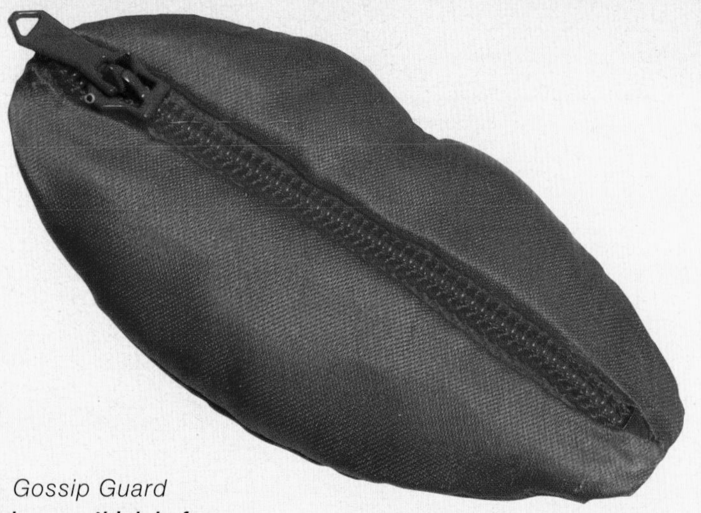

Gossip Guard

✱ Helps you think before you speak

In a free flow gossip situation, even the most socially upstanding of us can find that we have let mischievous cats out of their bags. The Gossip Guard helps reduce damaging prittle prattle by giving you that extra moment of reflection in which to check the impulse to start wagging your tongue. The device comprises a pair of lips in comfortable hygienic polyester, with a zipper to keep the mouth closed. When, and only when, the wearer is confident that his or her proposed act of speech is free from malice or unreliable information, the zip should be opened and the words released to the waiting audience.

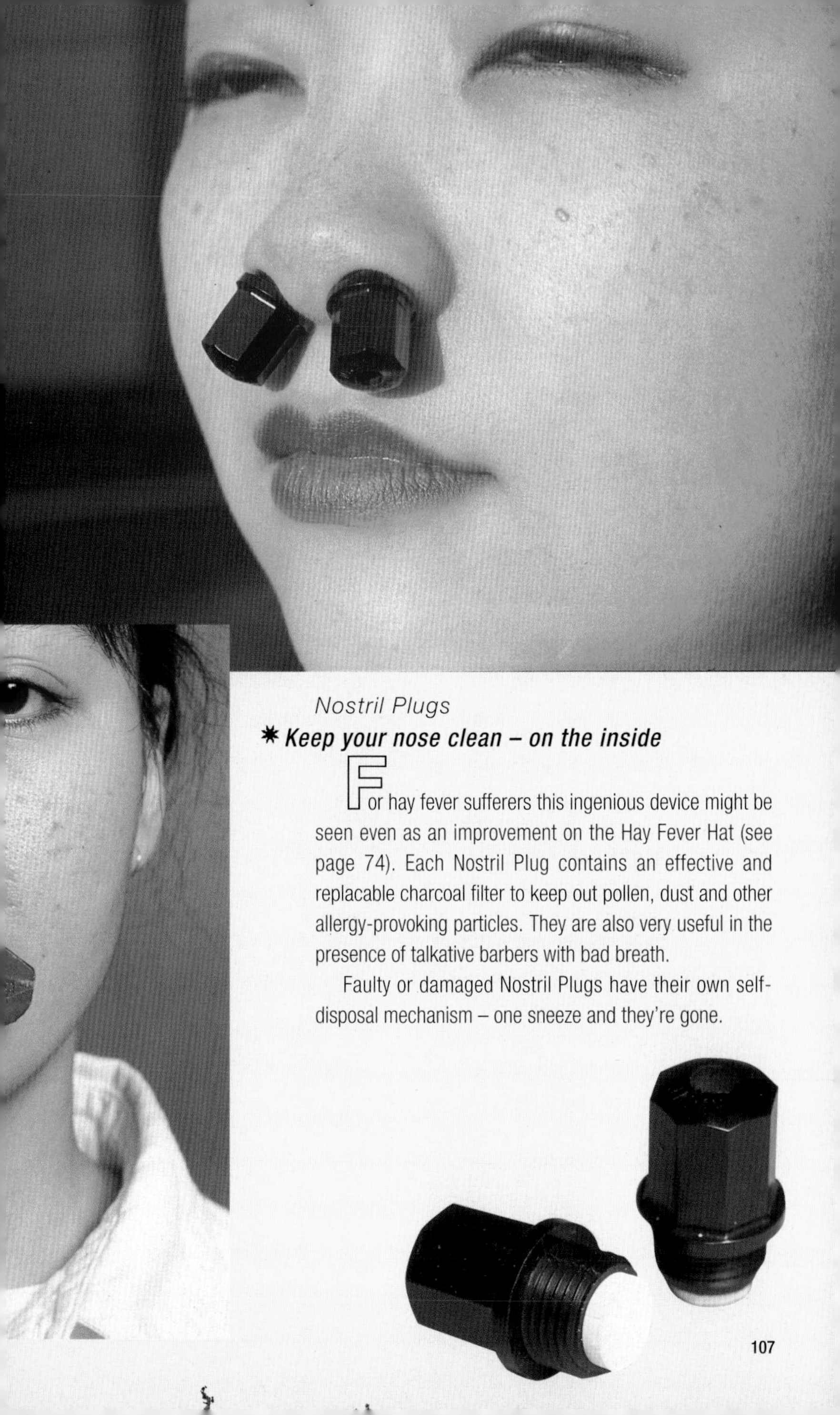

Nostril Plugs

✱ *Keep your nose clean – on the inside*

For hay fever sufferers this ingenious device might be seen even as an improvement on the Hay Fever Hat (see page 74). Each Nostril Plug contains an effective and replacable charcoal filter to keep out pollen, dust and other allergy-provoking particles. They are also very useful in the presence of talkative barbers with bad breath.

Faulty or damaged Nostril Plugs have their own self-disposal mechanism – one sneeze and they're gone.

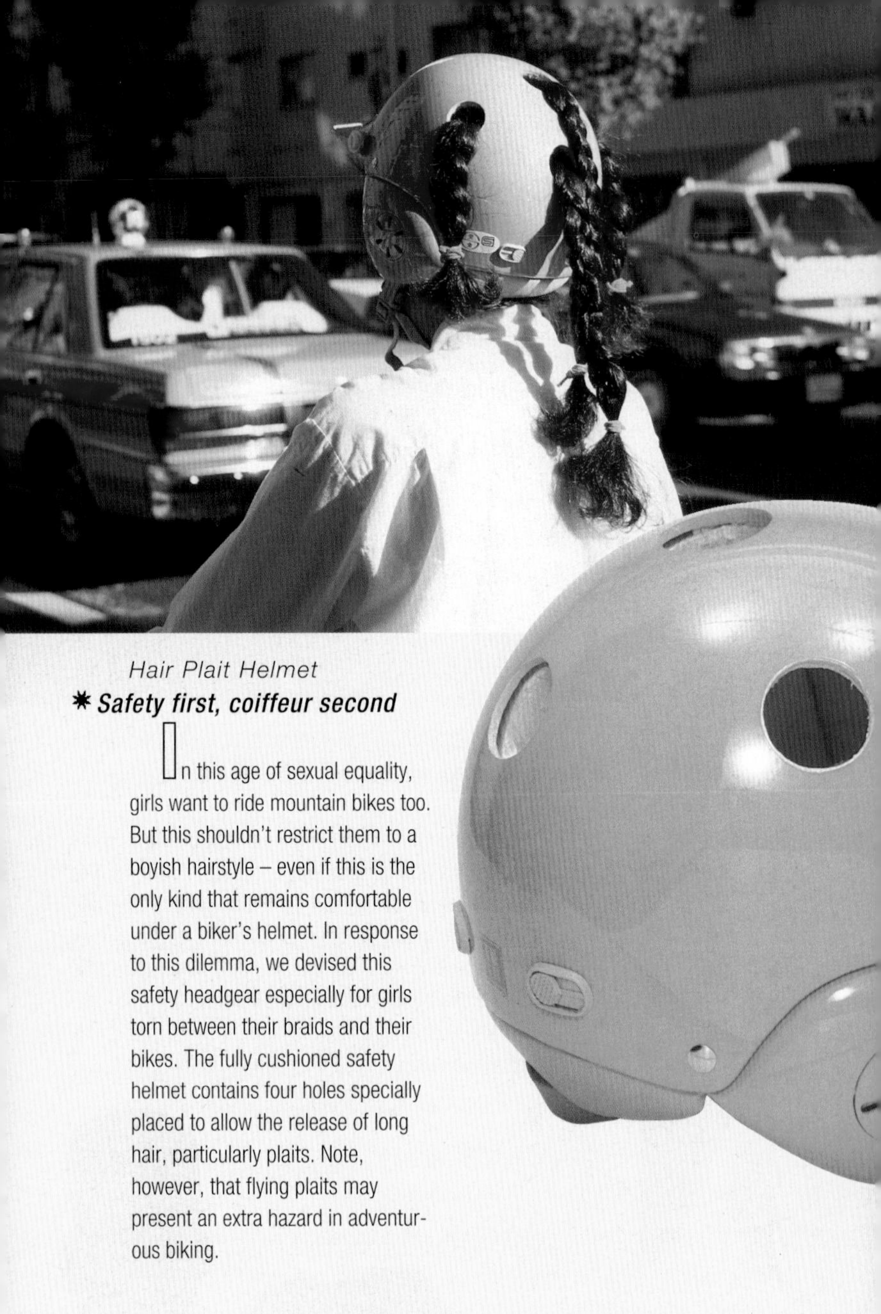

Hair Plait Helmet

✷ *Safety first, coiffeur second*

In this age of sexual equality, girls want to ride mountain bikes too. But this shouldn't restrict them to a boyish hairstyle – even if this is the only kind that remains comfortable under a biker's helmet. In response to this dilemma, we devised this safety headgear especially for girls torn between their braids and their bikes. The fully cushioned safety helmet contains four holes specially placed to allow the release of long hair, particularly plaits. Note, however, that flying plaits may present an extra hazard in adventurous biking.

Straight-in-the-pan Chopping Board

✱ Facilitates board–saucepan transfer of chopped ingredients

Next to the favourite knife, the chopping board is perhaps the cook's best friend. But it can seem more like a worst enemy when, on the journey to the saucepan, bits of chopped vegetable start to leave the board and relocate on the floor, work surface, and cooker. With this specially designed board such involuntary dispersal should no longer be a problem. The board can be placed so that the hole is directly over the saucepan, and the chopped ingredients are steered into the hole. Far too useful to be a real Chindogu? Maybe, except that cooks may find that, before it reaches the saucepan, as a method of ingredient dispersal the board-with-a-hole far outshines the original on which it was intended to improve.

珍道具

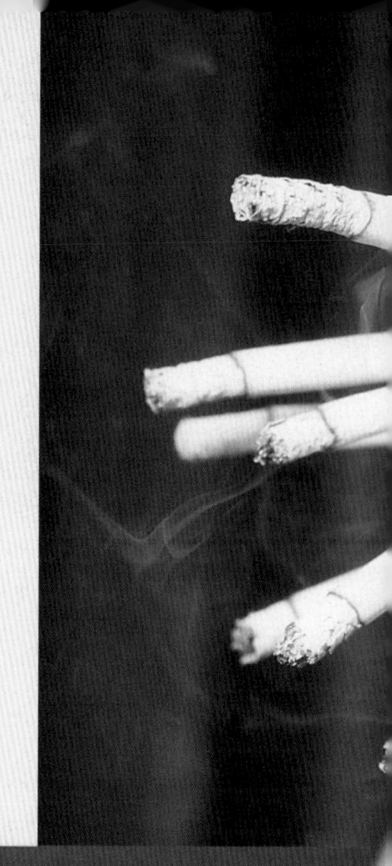

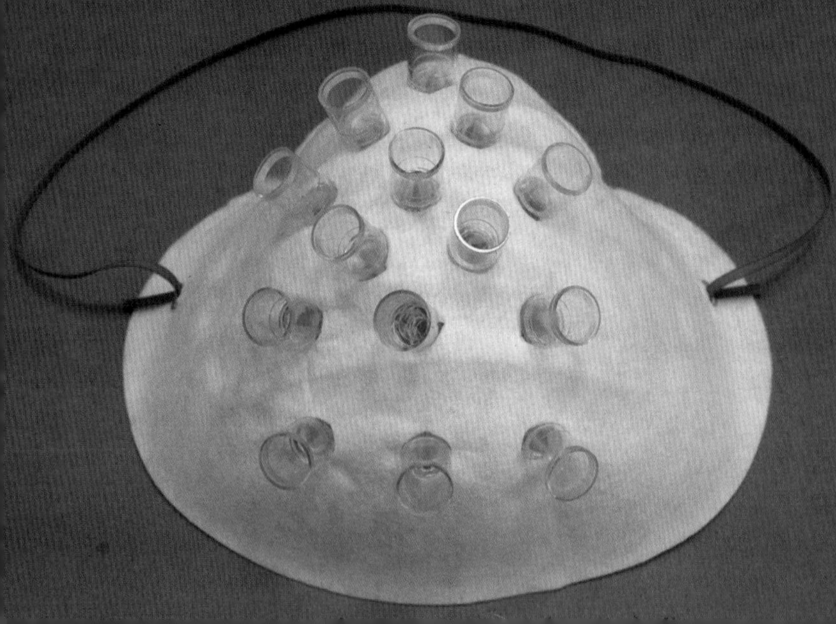

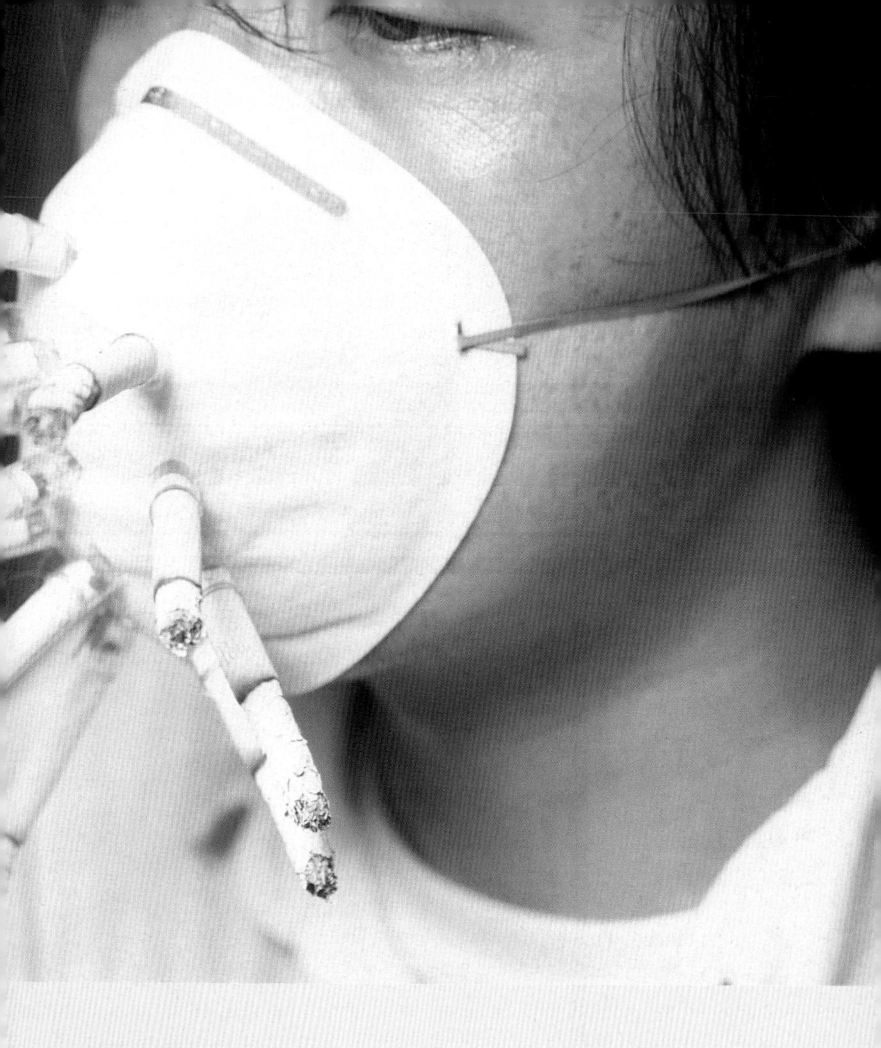

Heavy Smoker's Mask

✷ *Why smoke one when you can smoke fourteen?*

With no-smoking on public transport, non-smoking offices and non-smoking restaurants, it can be tough to get a cigarette in at all these days. So when you do find an acceptable smoker's niche, you may want to make up the lost ground. The Heavy Smoker's Mask allows you to smoke fourteen whole cigarettes in the time it would normally take to smoke one. So even a sixty-a-day smoker need only take five cigarette breaks to more than satisfy his or her habit.

Lawn Sandals

✷ *Always walk on the grass*

You can't beat the sensation of fresh green grass between your toes. But you can join it. The surface of the Lawn Sandals are covered with super-realistic artificial grass, to stimulate the soles and elevate the soul. You need never walk on tarmac or concrete again. Lawn sandals are fully washable, and guaranteed free of mud, moss and thistles.

Natto Cutters

✱ *For less messy brecky*

This device is a long overdue response to the problem posed by the famous Japanese breakfast dish, natto. This sweet, stringy, sticky foodstuff makes hot mozzarella look rigid by comparison. Lift natto with your chopsticks, and the strands just stretch and stretch. And when they do finally break, you can be sure they'll wrap themselves around your face, or cling to your clothes, leaving you with a persistent physical and olfactory reminder of what you had for breakfast.

Mounted on a one-size-fits-all bowl stand, the Natto Cutter allows you to select a manageable length of natto and snip it before it gets out of hand. Should bring peace of mind and reduced dry cleaning bills to natto lovers everywhere.

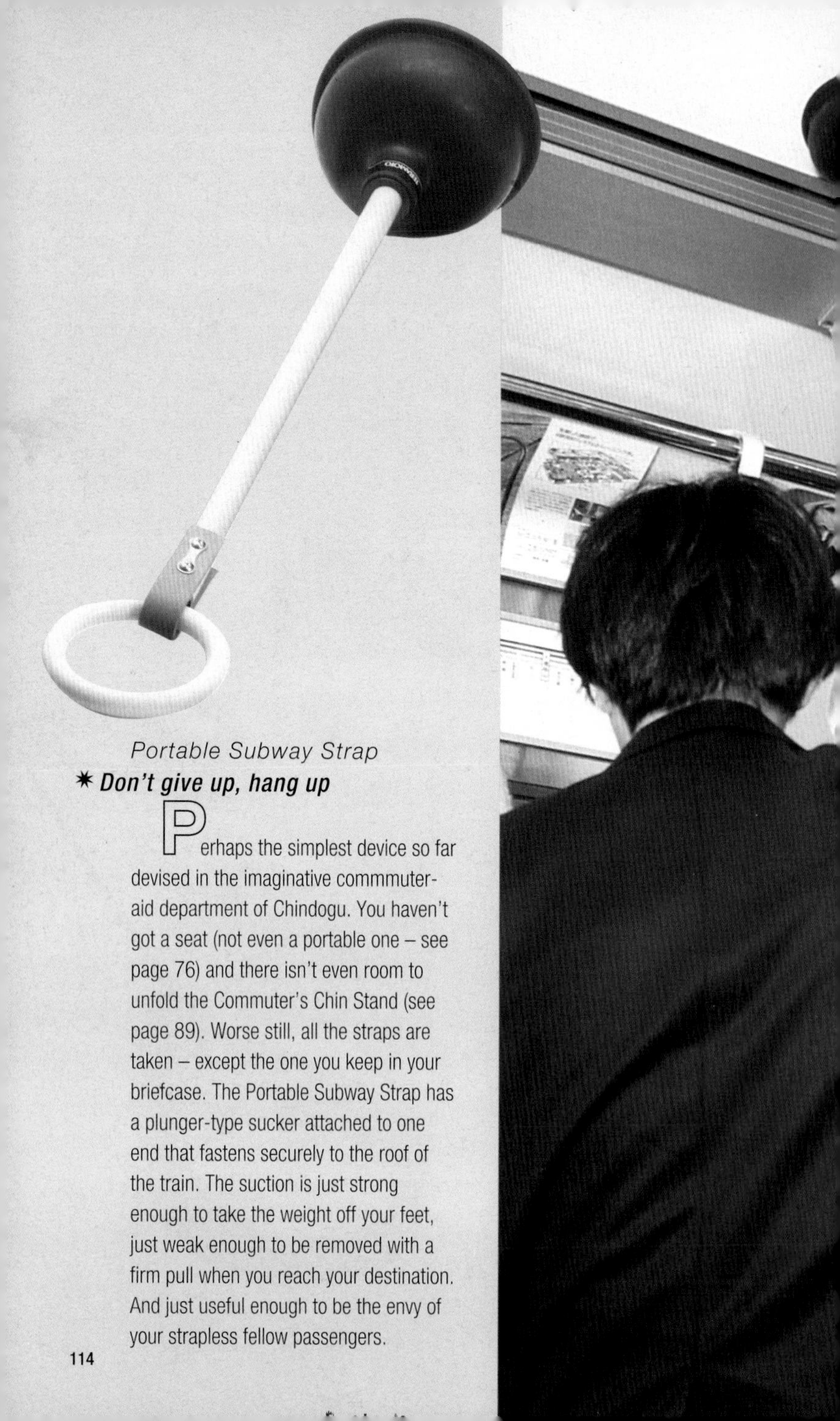

Portable Subway Strap

✱ *Don't give up, hang up*

Perhaps the simplest device so far devised in the imaginative commmuter-aid department of Chindogu. You haven't got a seat (not even a portable one – see page 76) and there isn't even room to unfold the Commuter's Chin Stand (see page 89). Worse still, all the straps are taken – except the one you keep in your briefcase. The Portable Subway Strap has a plunger-type sucker attached to one end that fastens securely to the roof of the train. The suction is just strong enough to take the weight off your feet, just weak enough to be removed with a firm pull when you reach your destination. And just useful enough to be the envy of your strapless fellow passengers.

11
23
日刊スポーツ
新
女王

Temporary Ladies' Room Converter

✷ *Stake your claim to the facilities*

In small restaurants and other public places where the best they can offer in the way of toilet facilities is a shared rest room, ladies in particular may find the arrangement unsatisfactory. In order to dissuade impatient men from coughing loudly outside the door and repeatedly trying the door handle, the Temporary Ladies' Room Converter will come in very useful. Just stick up the sign when the coast is clear and take all the time you like. And when you finally emerge, you can enjoy the sight of men wandering around in distracted confusion.

Multi-spicketed Watering Can

✷ You can't make it rain, but you can make it pour

Yet another application of the think in multiples branch of creative Chindoguing, the Multi-spicketed Watering Can greatly reduces irksome pot plant watering time. Nurseries and gardening centres can use it to reduce staff by seventy-five per cent. Note that pots should be carefully aligned, and also that greatest efficiency is achieved if the number of plants in your collection is divisible by four.

フジテレビを見て
泣くやつがあるか。
サンスウィミング
SUMMER BOY
BOOK'S

470-5011
MACMA
le Ponte
ASCOT

Golfer's Practice Umbrella

✷ *Work on your swing: swing on the way to work*

Men will be boys, and it's a common sight on the subway platform to see commuting golf enthusiasts practising their stroke with a furled umbrella. These besuited golfers tend not to impress however, bearing as they do a markedly greater resemblance to businessmen being eccentric with umbrellas than to sub-scratch handicap golf professionals. With a real driver head, the Golfer's Practice Umbrella is designed to improve both the performance, and the image in the eyes of critical fellow passengers, of these frustrated Nick Faldos. It is quite passably functional as an umbrella too, but is not recommended for use on a real golf course, as the user would look stupid.

4

Finger Combs

✱ *For more discreet personal grooming*

Nobody admires the vain individual who produces a comb in public and begins restyling their hair. Yet what can you do on the way to a date or important meeting when you can't get to the respectable privacy of a cloakroom? Miniature Finger Combs provide a discreet reassurance on all such occasions. Slip them on and pretend to be scratching your head over a difficult crossword clue, or just running your fingers through your hair in a carefree manner. No one need know that in fact you are removing unsightly tangles, or reinstating a lost parting.

Hairy Ego Booster

✱ ***Leave home without hair, but with confidence***

The bald man's pyschological dilemma is a complex one. To wear a wig or hairpiece would be to admit to friends and colleagues that your patial deficit bothers you. Yet wouldn't it be nice to look in the shaving mirror every morning and see a man with a lovely head of hair? Well now you can. The Hairy Ego Booster is an adjustable crop of real hair designed to stand between you and your shaving mirror to produce, by optical trickery, the illusion that you are in fact endowed with an impressive mane of thick locks. The result is reduced negative shaving mirror feedback and restored positive self image. You can leave the house each morning with a virile spring in your step. But be careful to avoid glancing at reflective or highly polished surfaces for the rest of the day.

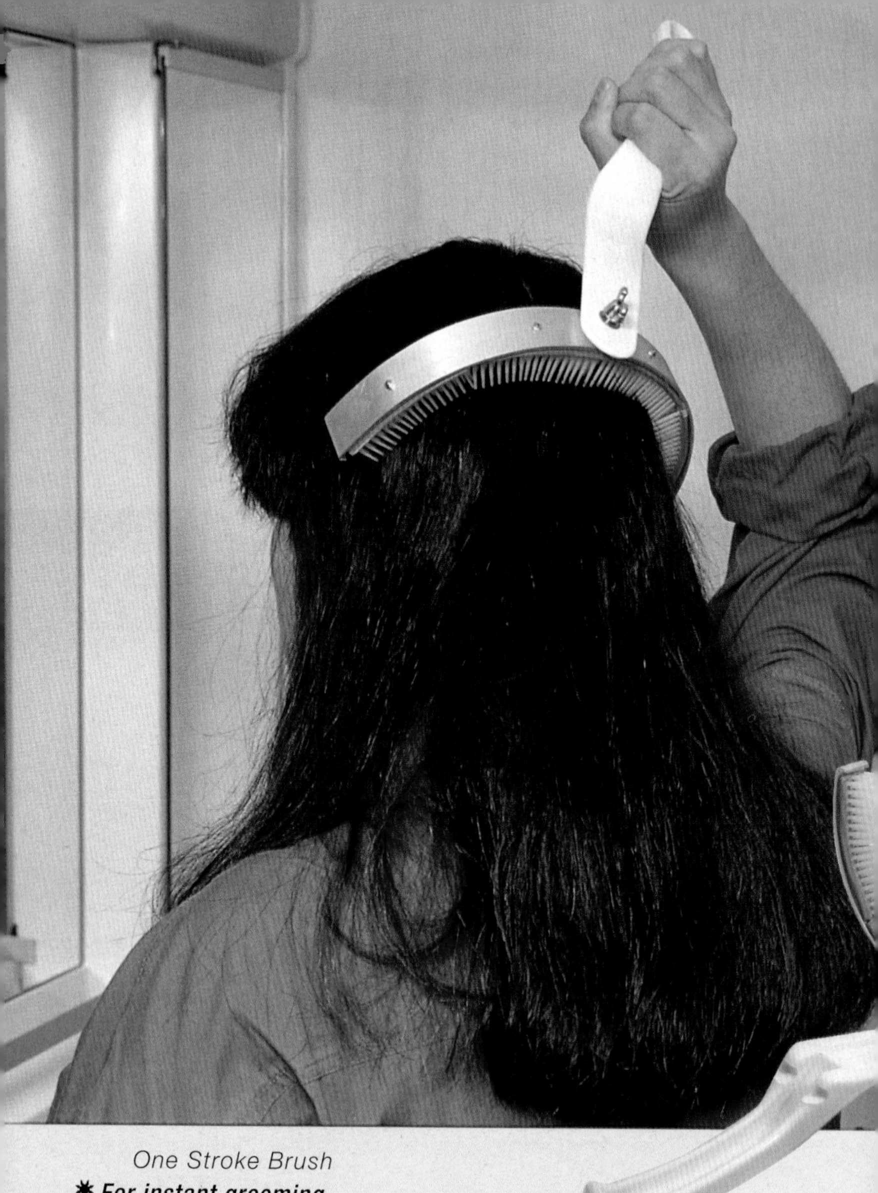

One Stroke Brush

✱ *For instant grooming*

Hair brushes have long been available in all sorts of shapes and sizes – except the obvious. So not before time we bring you . . . the hairbrush that actually fits your head. The scalp-contoured One Stroke Brush allows you to brush a whole head of hair (your own or a friend's) with a single firm stroke. Mousses and styling lotions can also be more quickly and evenly incorporated.

The Velcro Home Jogger

✱ *The economic alternative in home fitness*

Why pay for an expensive powered jogger's treadmill, when a little Chindogu ingenuity can give you a full jogging workout for a fraction of the price? Shod in special trainers soled with male 'hook' velcro, the exerciser jogs on the spot over a specially designed mat of female 'loop' velcro. The interaction between the two velcro surfaces produces sufficient resistance to require considerable physical effort. Ambitious indoor joggers may wish to consider carpeting their entire house with loop velcro to vary the available jogging landscape, or keep fit while they do the housework.

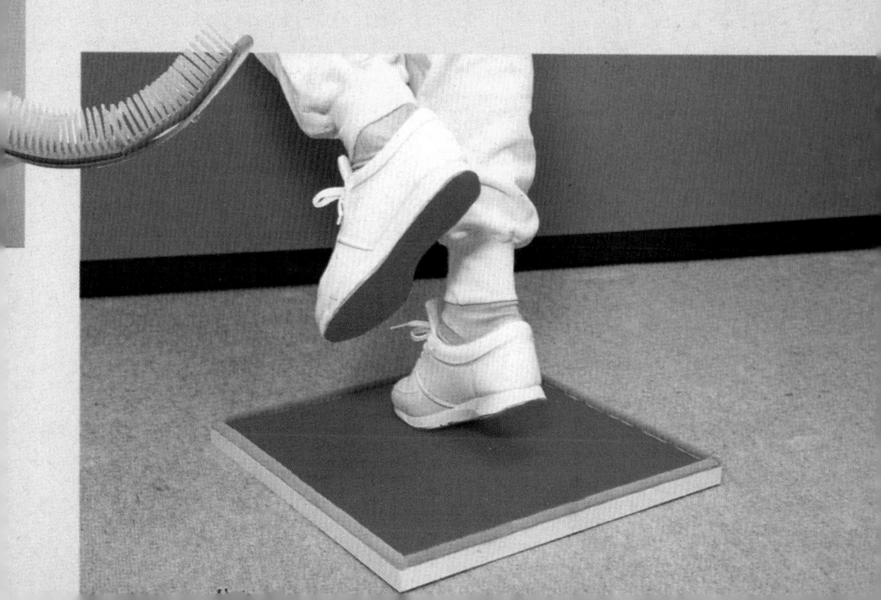

Vertigo Soothing Glasses

✷ ***Keep solid ground in sight***

Today's high-rise office blocks can strike terror into the hearts of the vertigo suffering salaryman. So what do you do when career advancement means a move to the fortieth floor? The Vertigo Soothing Glasses can't bring you back down to earth, but they can bring the earth up to eye level, for that comforting 'close-to-solid-ground' feeling.

Note: some vertigo sufferers may find the problem of not being able to see their own feet a high price to pay.

Golf Hoe

✱ *Drive out weeds, literally*

Another two-in-one Chindogu which allows the user to combine two of the most popular modern hobbies, golf and gardening. It takes a few well-timed swings to purge the lawn of a stubborn clump of weeds, and in so doing you will strengthen your golfing muscles and improve hand-eye co-ordination. The added bonus is that there is no need to replace divots (though this is not a habit to be carried on to the country club).

The Geta Shoe

✷ *Combined Occidental-Oriental footwear for men*

As East meets West in the business centres of the world, when it comes to formal business attire, the West has undoubtedly prevailed. The tunic and kimono have been supplanted by the pinstripe suit, and the traditional Geta sandal by the moccasin and the brogue. But homesick Japanese country boys, and Westerners who would like to sample the sensation of old style Oriental footwear, can take solace in the Geta shoe. As it keeps up Western appearances, it is sure to be acceptable in the business community. But the physical sensation of walking on these shoes is akin to hobbling along on a dirt road on the way back from the bathhouse.

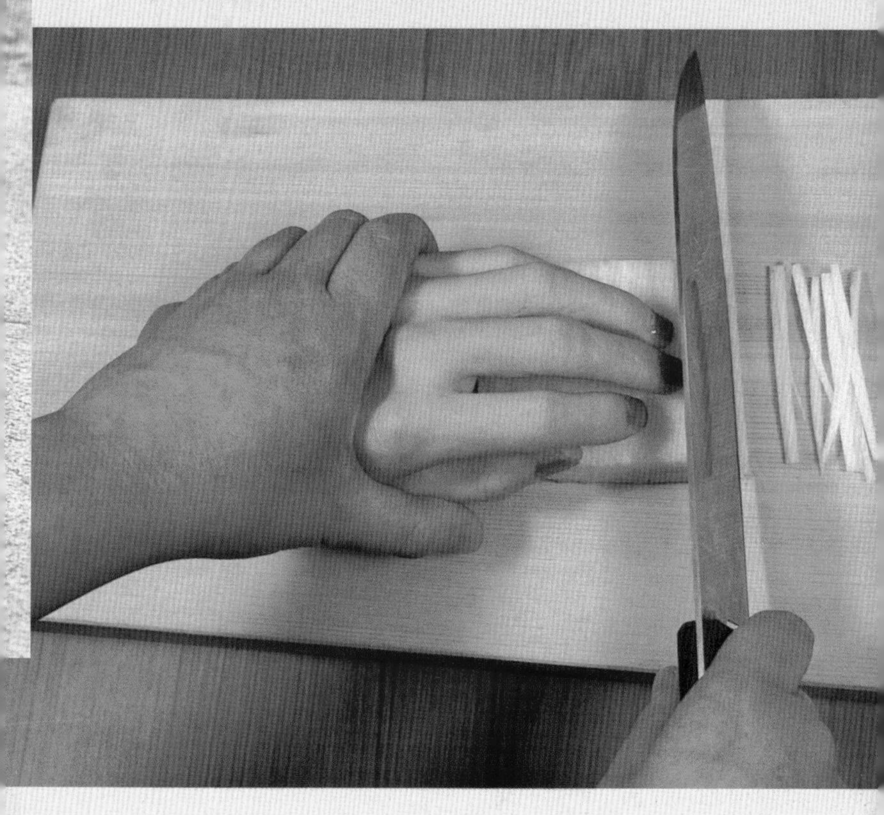

Finger Protector

✷ *For dangerous and dirty manual tasks*

This silicone substitute hand is designed to reduce the danger of cutting your own fingers when preparing food. Careless choppers can become carefree, slicing away at high speed, safe in the knowledge that damage done will be to a hand that doesn't mind. It won't recoil from dirty jobs either, so if you happen to be phobic about raw onions, fish scales, or the slippery vegetable debris that collects in the plug of the kitchen sink, then allow fearless silicone fingers to lend a hand.

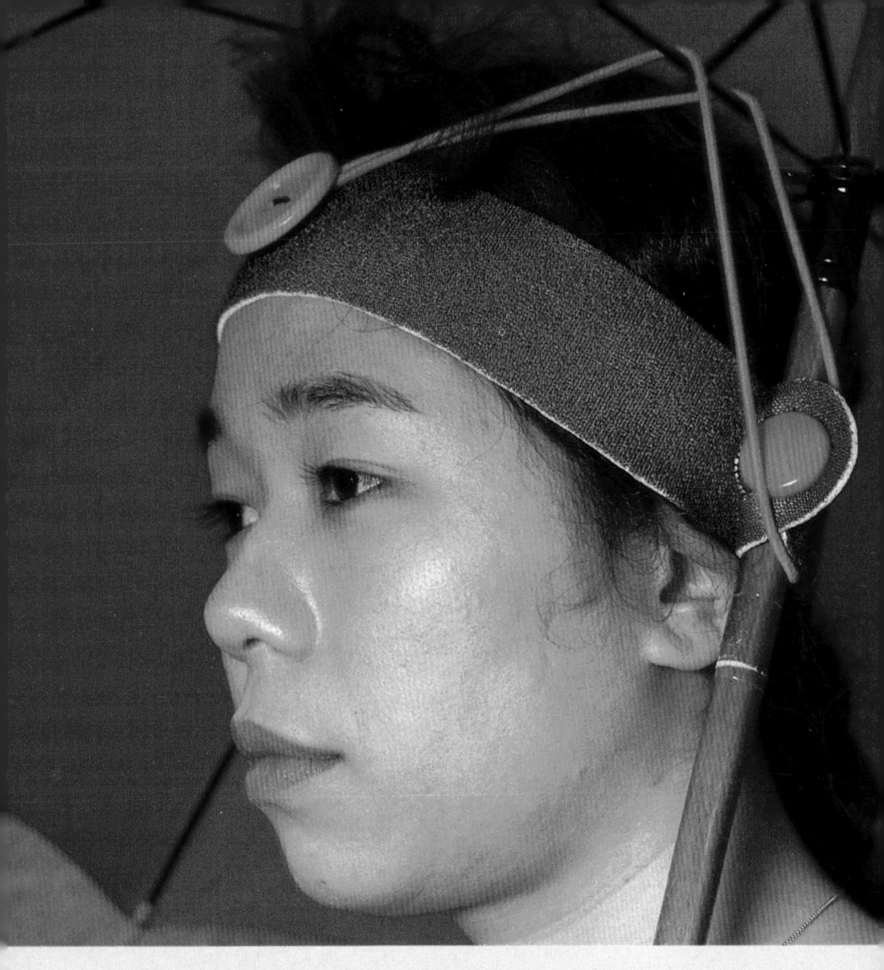

Umbrella Head Belt

✷ *When it starts to rain, use your head*

Another Chindogu to increase the range of activities available to the user during inclement weather, the Umbrella Head Belt can be used to adapt most conventional umbrellas for hands-free use. By strapping the umbrella stalk to your head, the belt makes both arms available for carrying shopping, suitcases or large pet dogs that don't like the rain. Some co-ordination and balance, not to mention neck strength, are required for successful deployment. When it's windy as well as wet, use of the Umbrella Head Belt is certainly not recommended.

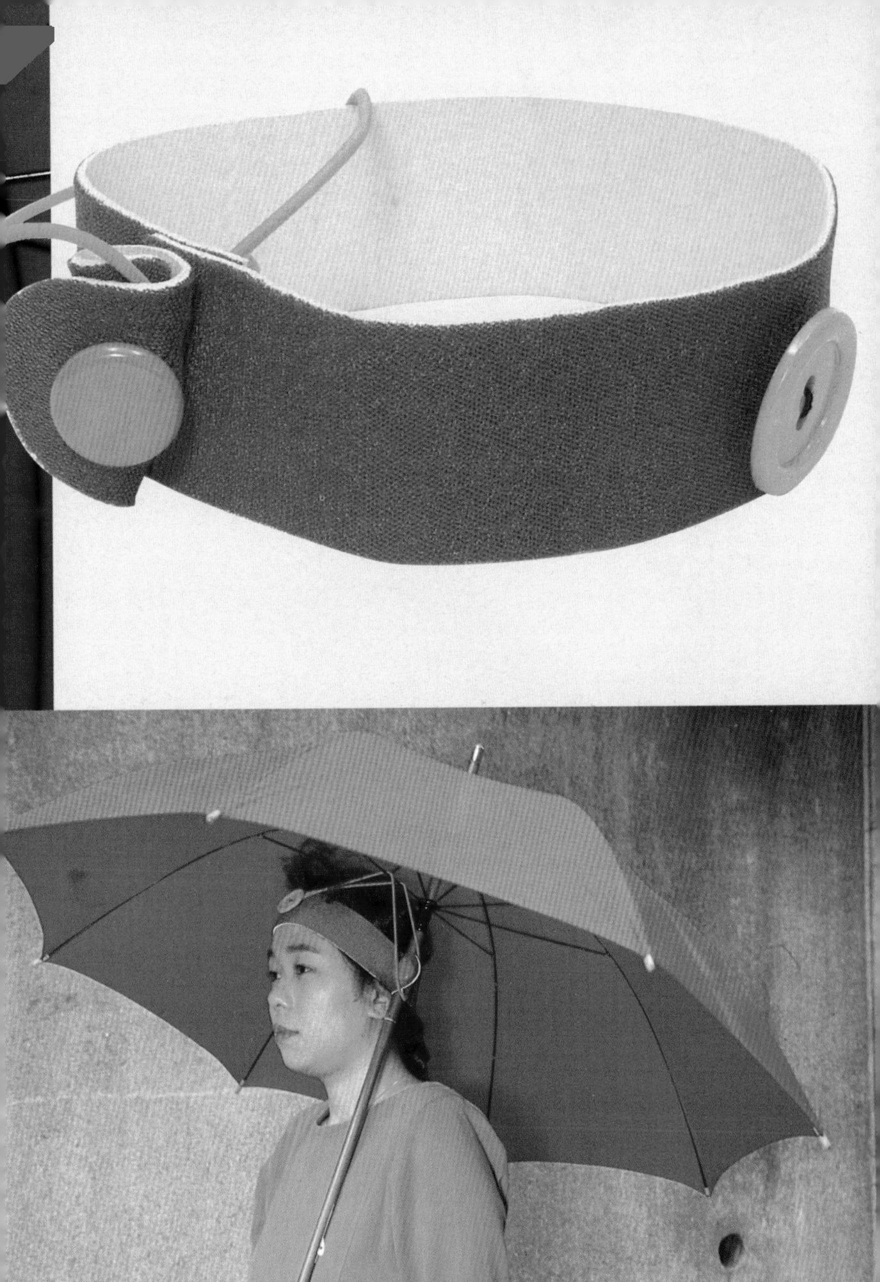

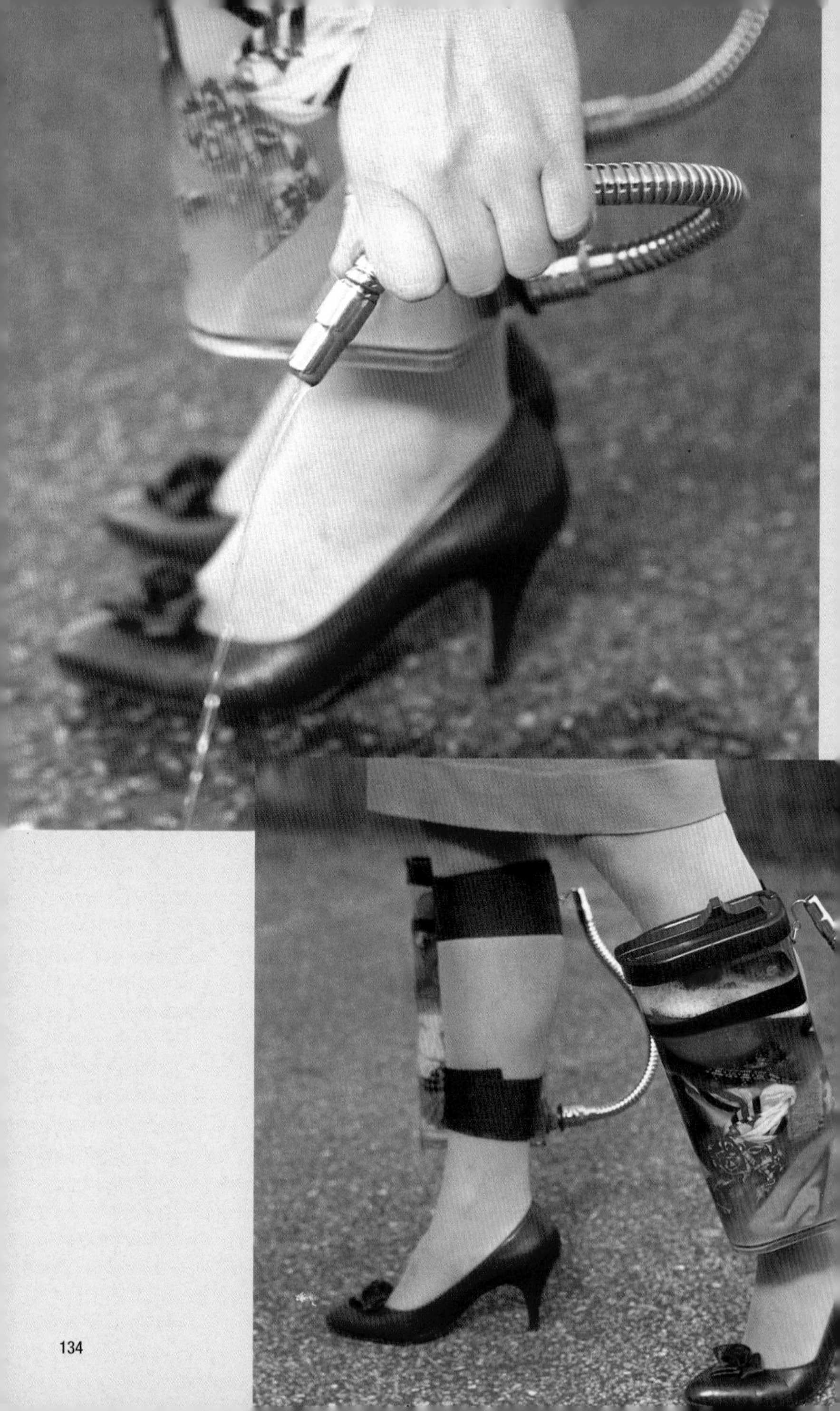

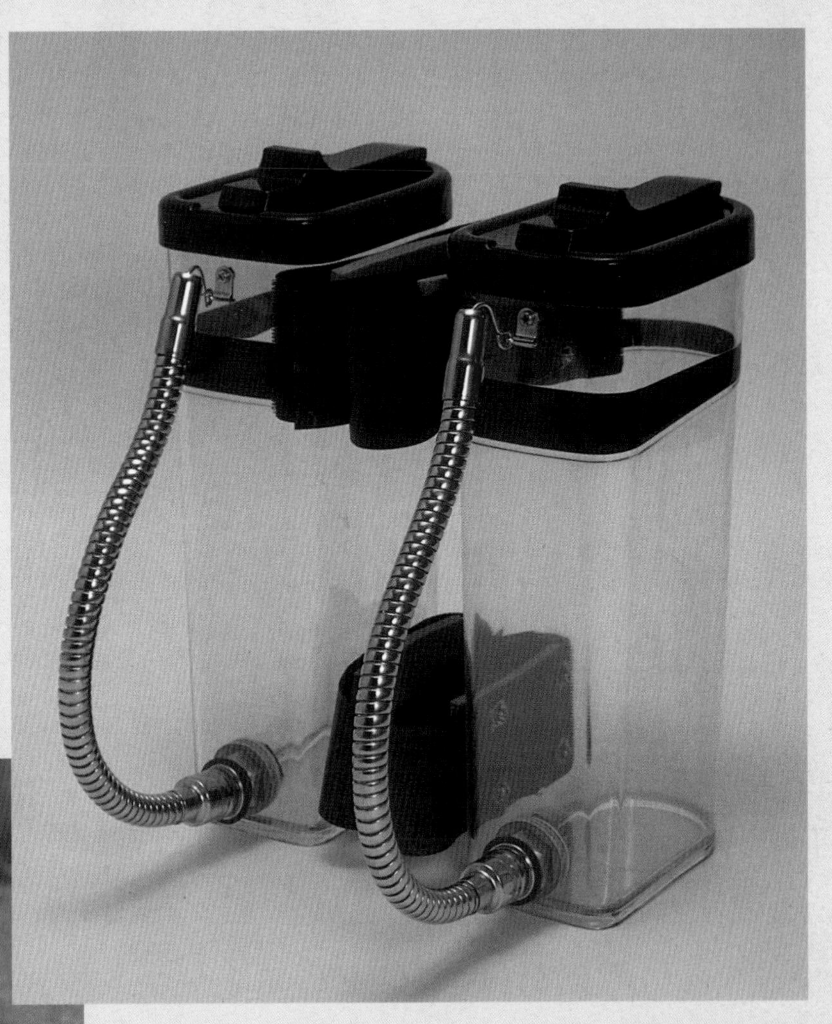

Walk 'n' Wash

✷ *Makes washing day a stroll in the park*

Some choices are hard, especially when guilt enters the frame. You'd like to go for a walk, but you ought to be doing the laundry. Well now you can do both. The Walk 'n' Wash comprises a pair of polythene tanks, one for each leg, with the capacity to hold two litres of laundry and water. So it's wash on the right leg, then rinse on the left (or vice versa, according to your personal preference). The weight of the tanks turns walking into a workout, making this a triple-barrelled Chindogu.

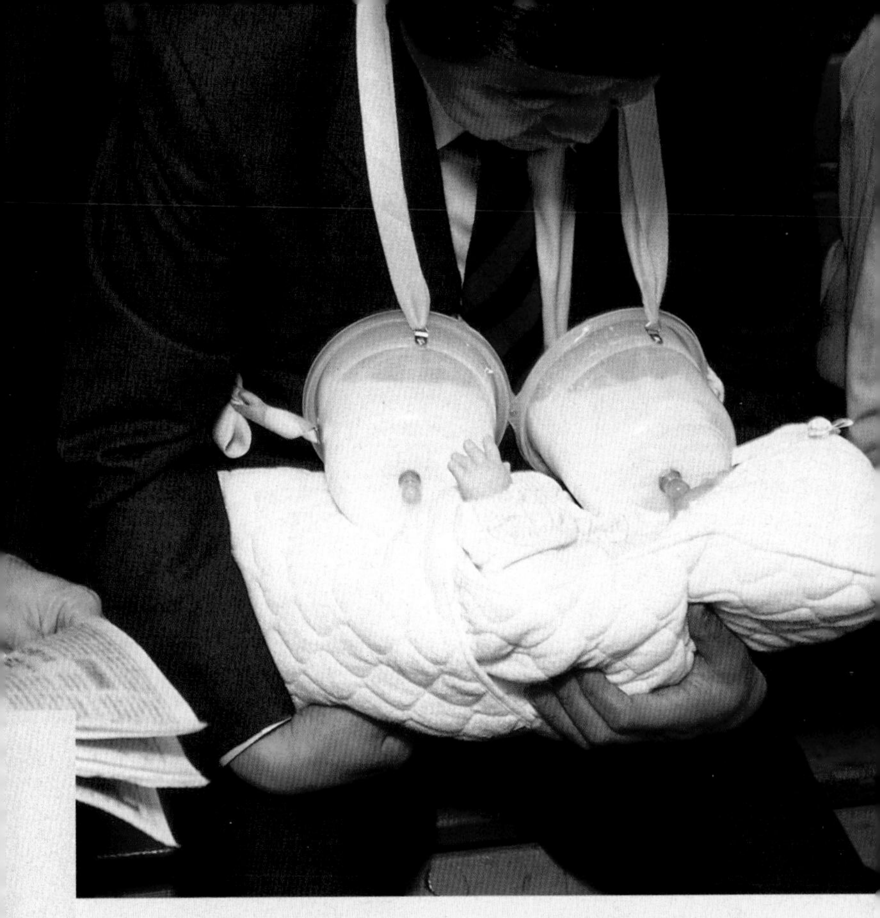

Daddy Nurser

✷ ***Lets Dad experience the joys of motherhood***

The division of work and responsibility within the modern family is constantly up for revision. Yet there is one task that still falls inevitably to the mother: breast-feeding the infants. This Chindogu challenges one of the last bastions of sexual inequality, and allows Father to experience the joy of nourishing his baby from his own body – almost. If this device was adopted on a large scale, then the benefits in terms of father–child bonding and the feminisation of the male could produce greater love and understanding on a global scale. Maybe.

Umbrella Drip Collector

✷ Protects floors from intrusive rainwater

Umbrellas are very good at keeping the rain off us when we are outdoors, but by a curious paradox they are also uncannily successful in bringing surprising quantities of rainwater indoors. The simple solution is the Umbrella Drip Collector – an inverted cup that attaches to the top of the umbrella. When the umbrella itself is inverted for temporary storage, the cup is the right way up, and well placed to catch the inevitable trickles. When you open the umbrella again, the collected water will run down the canopy and off into the gutter. (The old adage about not opening umbrellas indoors becomes particularly salient with this Chindogu.)

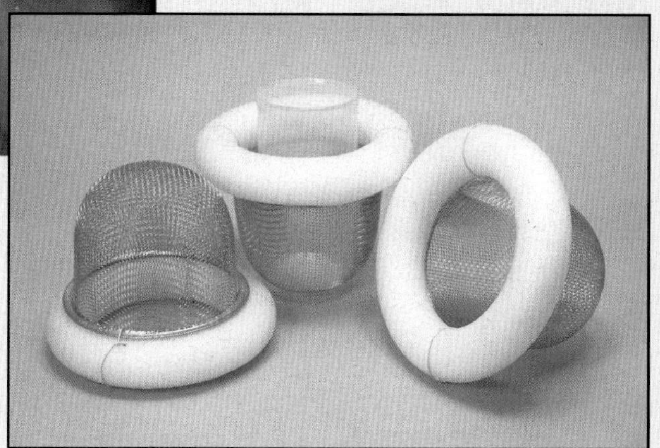

Hot Sake Bath Cup Holders

✷ *Makes sake drinking good clean fun*

Hot water is a resource too precious to waste on the unnecessary luxury of a long, lingering bath – which then goes down the plug hole. For greater energy efficiency, why not harness the heat of the bath to warm a few cups of sake? These sake cup-sized stainless steel sieves are suspended from miniature floating rubber rings, and are just deep enough to give maximum immersion to a near full cup of sake.

It is recommended that sake cups are placed in the holders after you have finished soaping. Careful bathers who do not splash too much may then enjoy a warm sake while they rinse.

Portable Zebra Crossing

✷ *The pedestrian's best friend*

The tyranny of the automobile makes life increasingly tough for eco-conscious pedestrians, and finding a safe place to cross can result in inconvenient diversions and wasted time. Now the pedestrian can fight back. When you've found the crossing point that best suits you, simply roll out the the Portable Zebra Crossing in front of you and cross confidently and in safety at your own pace.

Warning: on busy roads where there is no break in the oncoming traffic, attempting to roll out the Portable Zebra Crossing can be hazardous.

Drain Cleaning Boots

✷ *Makes light work of a heavy job*

Inadequately cleaned drains and gutters can harbour filth and disease. Cleaning them out has never been a pleasant task, or one you would want to get your face or hands too close to. Well now you don't have to, because this sturdy pair of rubber boots is fitted with a fork and trowel – all you need to clear debris and establish a free flowing drain again. Once it was the least favourite on the list of family chores, but now they'll be queuing up to give the Drain Cleaning Boots a go.

Subway Sleeper's Screen

✷ *For more efficient sleep stealing*

The extra few minutes of sleep to be gained on the subway to and from work can be a valuable boost to your natural bedtime repose, but at what cost to personal dignity? Exhausted commuters are frequently seen slumped in their seats, legs apart, mouths open, snoring loudly. The Subway Sleeper's Screen addresses several anxieties of the underground snoozer, promoting a deeper, more refreshing sleep. Firstly it conceals his or her identity, hides an open mouth, and even goes some way towards muffling the sound of snoring. Secondly it masks the area that would be in view if the sleeper has fallen asleep with legs wide apart. Thirdly, the screen is emblazoned with the name of the station where you need to alight, so helpful (and appreciative) fellow passengers can wake you up in time.

Training High Heels

✱ *Smooths the transition to adult female footwear*

The graduation to high heels is something a young girl looks forward to as much as make-up and boyfriends. But when the big day comes, it can be frightening to mount those tottering towers and learn to walk all over again. An uncertain or lacklustre performance defeats the whole purpose of these elegant accessories. For beginnners, therefore, we recommend the Training High Heels. With miniature stabilising wheels attached to the pointy heels of either shoe, you can walk without fear of tipping over. When confidence and poise have been achieved, the wheels are removed and the young graduate can go solo.

Ten-in-one Gardening Tool

✷ *The amateur gardener's do-it-all*

The multi-faceted Swiss Army penknife is one of the great life-enhancers of modern times. By way of a tribute to the genius of this invention, Chindogu has moved the concept forward, and up in scale. The result is the Ten-in-one Gardening Tool, incorporating two sizes of shovel, a hoe, a blade, a saw, a set of pruning shears and several other gardening implements, all in one (moderately) compact unit. No more endless trips to and from the shed, and when the day's work is done, you won't have to remember all the tools you've scattered around the garden. However, you may want to think twice about lending your hoe to a needy neighbour, as everything else will have to go with it.

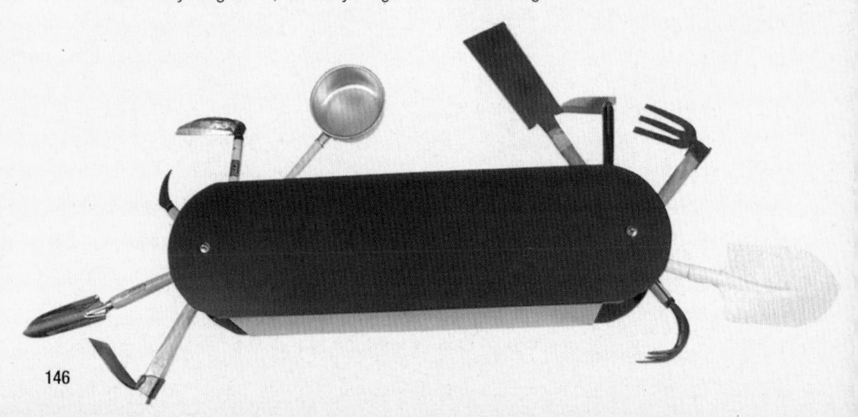

Head Handle for Shoulderbags

✱ *For a more balanced approach to shopping*

In just a morning tramping round the shops, a heavy handbag can play havoc with our physical equilibrium, causing shoulder pain, and back ache. So give the shoulder a rest: use the Head Handle to convert your handbag to a headbag. An attractive gold chain connected to a deceptively sturdy headband promotes more symmetrical distribution of weight. When you start to get a pain in the neck, you can revert to the traditional shoulder position.

Lawn-mowing Sandals

✷ *Labour-saving garden footwear for the supremely idle*

The prevalence of back yard jungles is a testament to the reluctance of the urban male to get round to mowing the lawn. The Lawn-mowing Sandals are designed to lure the reluctant mower back into the garden. Each has a miniature scythe protruding from the heel, so that even those who prefer a leisurely stroll to heaving a lawn mower up and down the yard can still get the grass cut. The steady shuffle of the truly indolent is more effective than vigorous striding up and down, which can have a tendency to take the blades over the top of the grass.

Outdoor Loo Seat

✷ *Brings the great outdoors to the smallest room*

For many people, lavatory time is leisure time, time for meditation and making peace with a hostile world. There's a lot to be said for improving the quality of this valuable time, and what better way to do it than by bringing nature even closer in this most natural of communions? The Outdoor Loo Seat is covered with convincing artificial grass that puts you instantly in touch with the wilderness. Piped birdsong and running water sounds will further enhance the experience, and those looking for true authenticity might consider sprinkling a few ants amongst the blades of plastic grass.

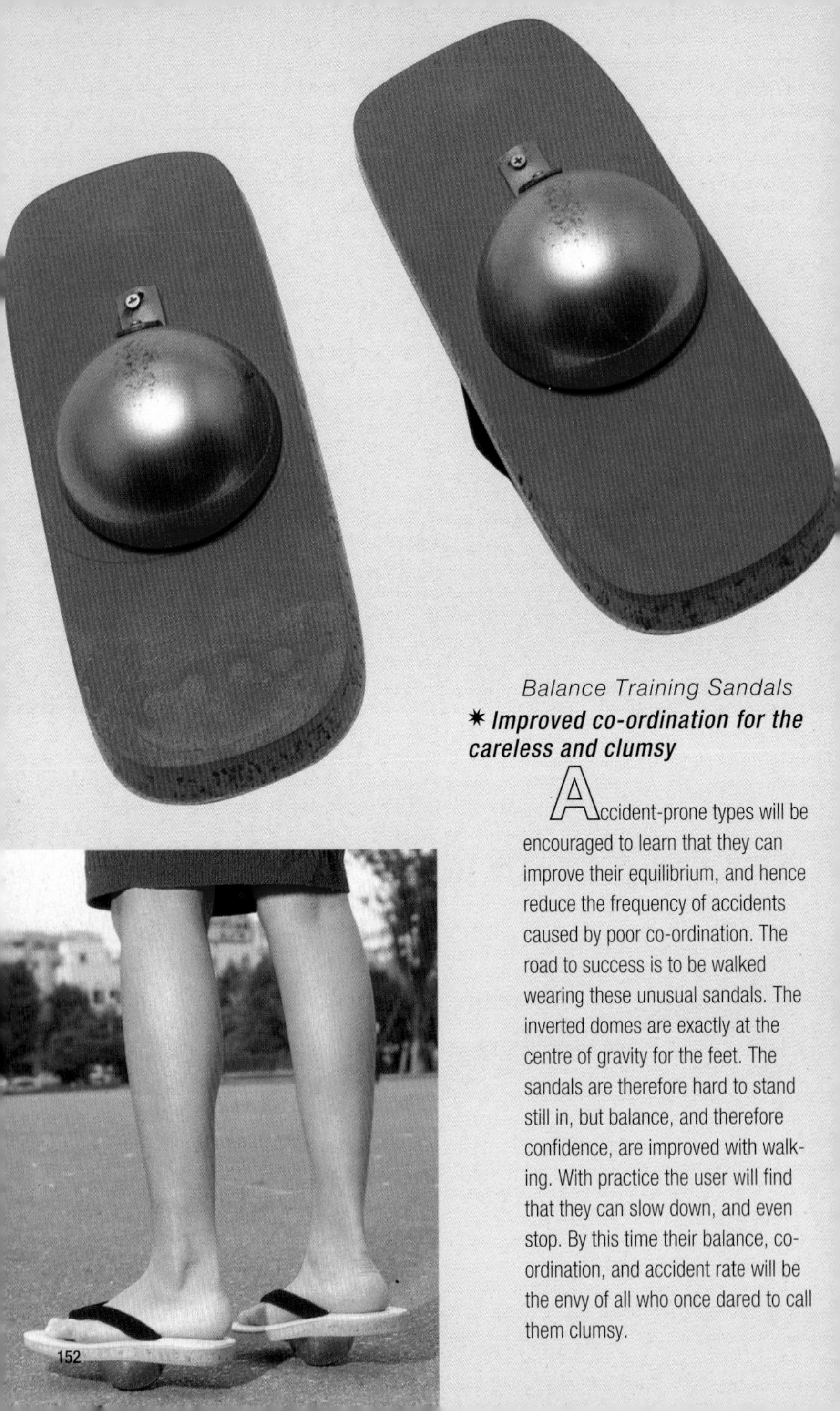

Balance Training Sandals

✷ Improved co-ordination for the careless and clumsy

Accident-prone types will be encouraged to learn that they can improve their equilibrium, and hence reduce the frequency of accidents caused by poor co-ordination. The road to success is to be walked wearing these unusual sandals. The inverted domes are exactly at the centre of gravity for the feet. The sandals are therefore hard to stand still in, but balance, and therefore confidence, are improved with walking. With practice the user will find that they can slow down, and even stop. By this time their balance, co-ordination, and accident rate will be the envy of all who once dared to call them clumsy.

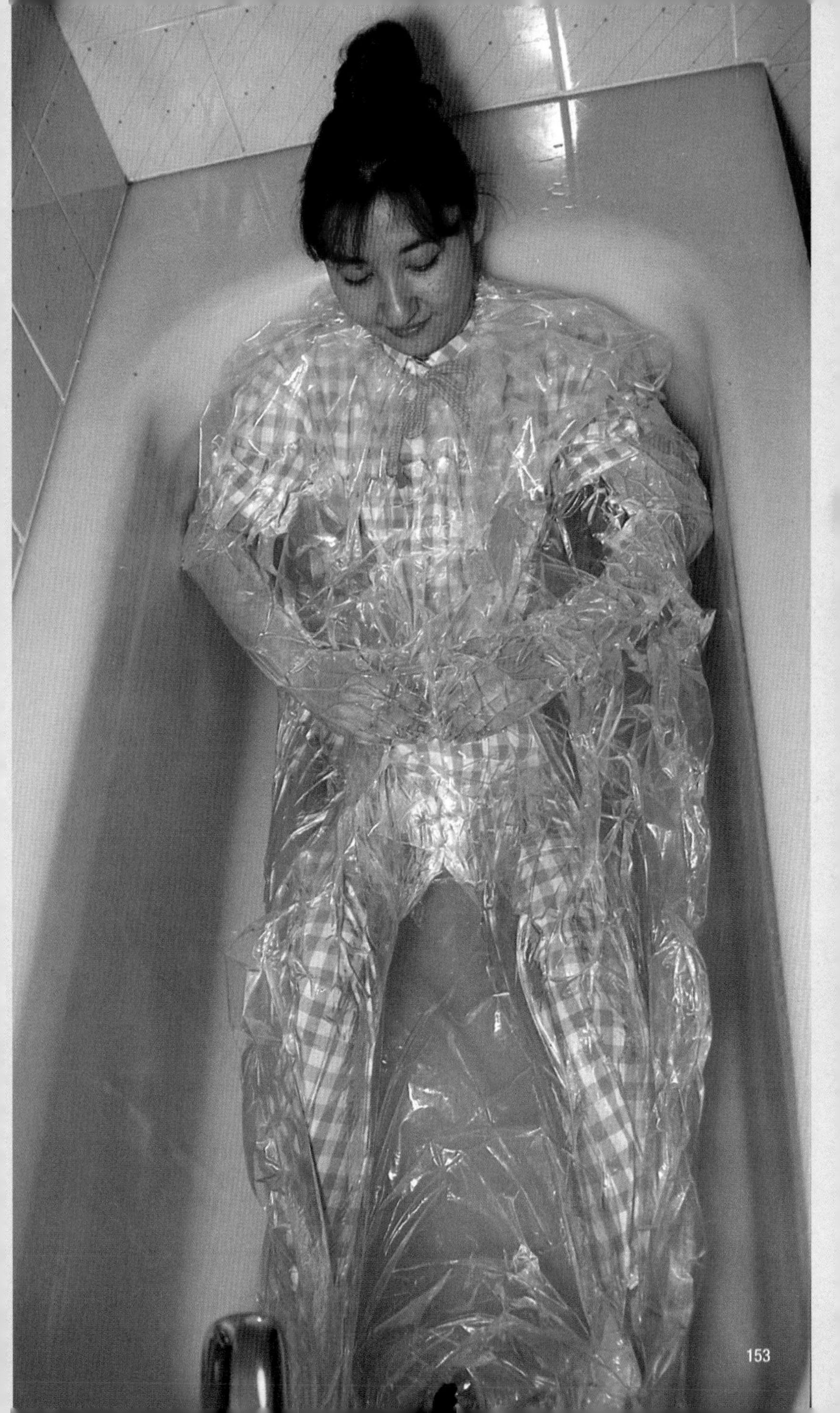

Hydrophobe's Bath Body Suit

✱ *Take a bath without getting wet*

The dry shave has already saved millions of man hours the world over. The dry bath is an even more radical concept, with the potential for even greater impact on the way we live. The benefits are considerable. No need to undress. No need to dry, talc up and redress. You can experience all the warmth, comfort and therapeutic relaxation of the conventional bath, without the wetness. You can then get out of the bath, surrounding yourself with the cool air of the bathroom, without the unpleasant chill factor that so often spoils the end of bath experience for the wet and naked.

Such is the pleasure and convenience of the dry bath, that wearers of the body suit may be inclined to 'soak' for very long periods. This is to be discouraged, as resulting excessive perspiration may require the remedy of a conventional wet bath.

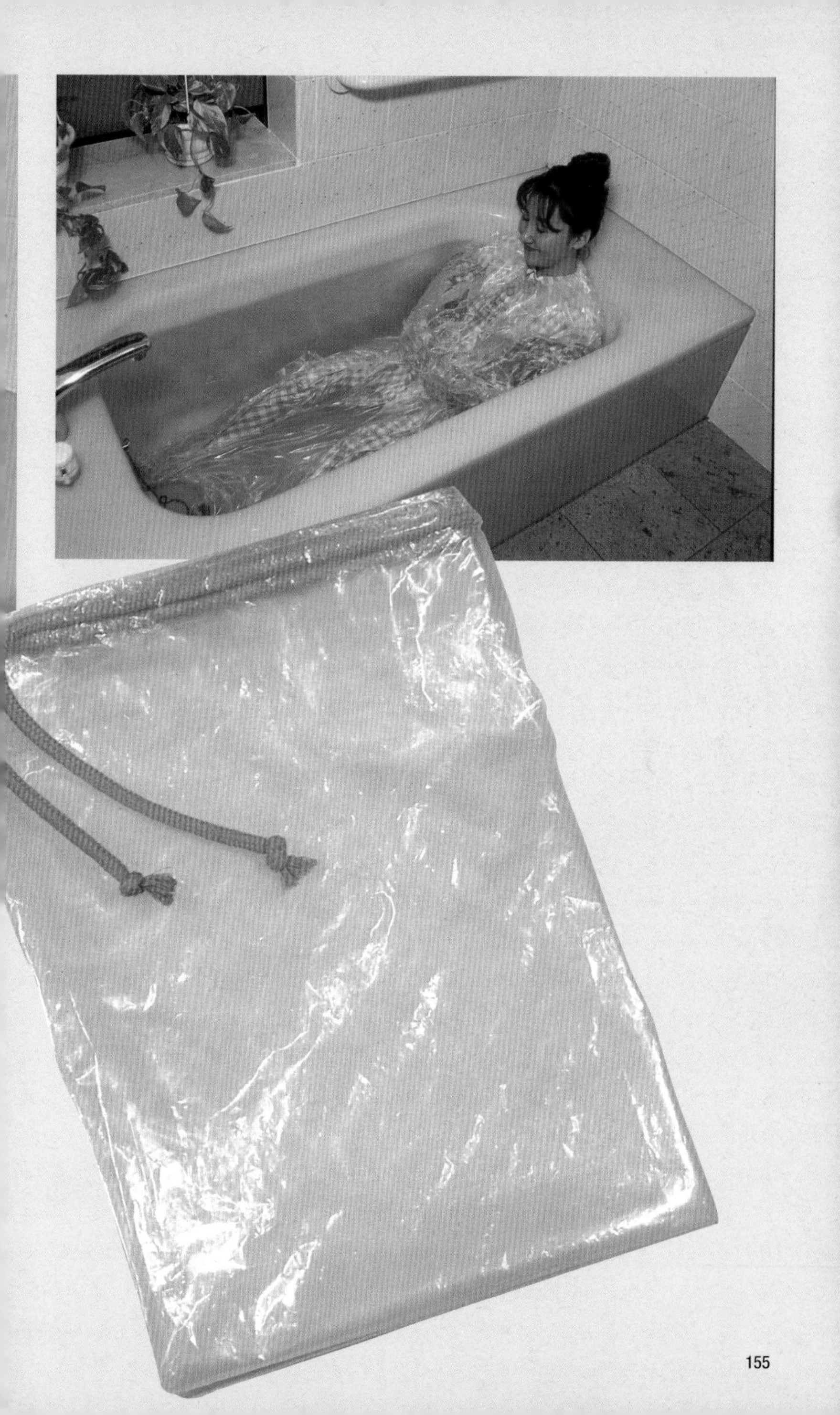

11/23
ただいま眠っております。
お手数ですが下記の駅で
起こしてください。
西荻窪
多謝

Commuter's Helmet

✷ *For secure subway snoozing*

Another Chindogu to aid the comfort and repose of weary commuters, and arguably the best. Like the Subway Sleeper's Screen (see page 143), the Commuter's Helmet sports a message to fellow travellers, reading, 'I'm having a short nap. Could you please wake me up when I reach the stop printed below. Many Thanks.' But since it depends entirely on the co-operation of fellow passengers for success, this Chindogu has also been designed to maximise their finer communal instincts and sense of goodwill. The suction pad on the back of the helmet keeps the head firmly in place, thus preventing the sleeper's head from lolling intrusively on the shoulders or laps of his or her neighbours. This courtesy will no doubt be appreciated, and the reciprocal favour of a timely awakening is more likely to be achieved.

Drink De-carbonator

✷ *Curbs troublesome bubbles*

When daily after-work drinking is an offer you can't refuse, the strain on the stomach of the congenial salaryman can be too much to bear. Apart from alcohol itself, the gas in beer and carbonated mixer drinks is perhaps the prime culprit in professional dispepsia. Clearly we can't dispense with the alcohol; but we can, at last, dispense with the gas.

When a bubbly beverage is placed in the utility cup of the Drink De-carbonator, the switch activates high-powered vibrations which awaken the gas molecules and hasten their natural upward departure. Beers and gin and tonics are quickly rendered flat, and thereby less irritating to sensitive stomachs. Alcoholic drinks can be consumed in greater quantities, and the drinker can then focus his attention solely on the problem of inebriation.

Automobile Parasol

✷ *A cool car for a cool head*

On a sunny day, we'd all like a cool shady parking spot. But we rarely succeed in finding one, and the result is a hot car, steaming kids, melting pets – and inevitably a scorching temper on the part of the driver.

So if you can't park the car in the shade, park the shade on the car. The automobile parasol is over a metre and a half in diameter, and casts a big enough shadow to accommodate even a large family saloon. Can also be used in very slow traffic jams.

The International Chindogu Society welcomes new members. If, after reading this book and acquainting yourself with the principles of Chindogu, you feel that you are ICS material and would like to take the next step, please fill out the short form below and send it to the Chindogu Society office appropriate to your locality.

The requirements for ICS membership are a sincere interest in and understanding of Chindogu and the contribution of one idea (with illustration) which we may build, photograph and include in future Chindogu literature. Please include a G-class stamp (from within the US and Canada) or an international reply coupon for the equivalent of one British pound so that we may send applicants a membership kit.

Thank you for your interest and we hope that the Chindogu International Society will be the first step on your journey towards a fuller and more productive life.

International Chindogu Society
Overseas Operation
18433 Hatteras Street, Suite 507
Tarzana Ca 91356 US

Chindogu Society Home Office
Flower Building, 5th Floor
1-25-3 Hongo
Bunkyo-ku, Tokyo 114 Japan

International Chindogu Society
c/o HarperCollins Publishers
77–85 Fulham Palace Road
Hammersmith , London W6 8JB UK

Personal Information

Name: . . . . .Age: . . . . . . .Sex: . . . .

Occupation: . . . .
Street Address: . . . .
City, State, Country: . . . .
Post Code: . . . .
Home Phone: . . . .

Comments or Questions . . . .
. . . .
. . . .
. . . .
. . . . . .
Your Chindogu Suggestion: . . . .
. . . .
. . . . . .
. . . .
. . . .
Name: . . . .
Explanation: . . . .
. . . .
. . . .

珍道具